Anil K. Sharma, Girish Kumar Gupta, Mukesh Yadav (Eds.)
Medical Microbiology

Also of Interest

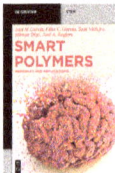

Smart Polymers.
Principles and Applications
García JM, García FC, Reglero Ruiz, Vallejos, Trigo-López, 2022
ISBN 978-1-5015-2240-6, e-ISBN 978-1-5015-2246-8

Drug Delivery Technology.
Herbal Bioenhancers in Pharmaceuticals
Pingale, 2022
ISBN 978-3-11-074679-2, e-ISBN 978-3-11-074680-8

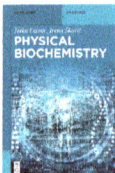

Physical Biochemistry
Dumić, Škorić, 2022
ISBN 978-3-11-056389-4, e-ISBN 978-3-11-056391-7

Microbiology of Food Quality.
Challenges in Food Production and Distribution During
and After the Pandemics
Hakalehto (Ed.), 2021
ISBN 978-3-11-072492-9, e-ISBN 978-3-11-072496-7

Food Science and Technology.
Trends and Future Prospects
Ijabadeniyi (Ed.), 2020
ISBN 978-3-11-066745-5, e-ISBN 978-3-11-066746-2

Medical Microbiology

Edited by
Anil K. Sharma, Girish Kumar Gupta, Mukesh Yadav

DE GRUYTER

Editors

Prof. Dr. Anil K. Sharma
Department of Biotechnology
Engineering Block-II
M.M. (Deemed to be University)
Mullana, Ambala 133207
Haryana
India
Tel: +91-8059777758
Email: anibiotech18@gmail.com;
sharma.biotech@gmail.com

Dr. Mukesh Yadav
Department of Biotechnology
Engineering Block-II
M.M. (Deemed to be University)
Mullana, Ambala 133207
Haryana
India
Email: mukeshyadav7@gmail.com

Dr. Girish Kumar Gupta
Department of Pharmaceutical Chemistry
Sri Sai College of Pharmacy
Badhani, Pathankot
Punjab
India
and
Director of Research and Development
Sri Sai Group of Institutes
Badhani, Pathankot 145001
Punjab
India
Tel: +91-7206930164
Email: girish_pharmacist92@rediffmail.com

ISBN 978-3-11-051764-4
e-ISBN (PDF) 978-3-11-051773-6
e-ISBN (EPUB) 978-3-11-051774-3

Library of Congress Control Number: 2021950683

Bibliographic information published by the Deutsche Nationalbibliothek
The Deutsche Nationalbibliothek lists this publication in the Deutsche Nationalbibliografie;
detailed bibliographic data are available on the Internet at http://dnb.dnb.de.

© 2022 Walter de Gruyter GmbH, Berlin/Boston
Cover image: designsstock/istock/Getty Images Plus
Typesetting: Integra Software Services Pvt. Ltd.
Printing and binding: CPI books GmbH, Leck

www.degruyter.com

Contents

About the editors

Prof. Anik K. Sharma is having 25 years of research experience including industrial R&D, teaching and administration. Currently, he is working as a Professor and Head in the Department of Biotechnology at Maharishi Markandeshwar (MM) (Deemed to be University) Mullana-Ambala, Haryana, India, for past 11 years (2010–till date). Before this assignment, he worked as a Senior Research Specialist in Health Sciences (2008–2010) and a Post-Doctoral Research Fellow (Molecular biology) (2003–2010) at the Microbiology and Immunology Department at UIC College of Medicine Chicago, IL, USA. Dr. Sharma has worked in diverse scientific fields ranging from molecular biology, cancer biology, antimicrobial drug resistance, nanomedicines to the development of microbial strains for remediation of heavy metals and pesticides. His contributions to medical biotechnology and industrial cum pharmaceutical biotechnology have been greatly acknowledged. Dr. Sharma has published 155 articles in peer-reviewed, high-impact journals of international repute with some of them in *Journal of Biological Chemistry* (ASBMB, USA, impact factor ~4.57), *Plant Biotechnology Journal* (impact factor 8.4), *Seminars in Cancer Biology* (Elsevier, impact factor ~15.7), *Anti-Cancer Agents in Medicinal Chemistry* (impact factor 2.7), *Immunology Cell Biology* (Nature Publishing Group, impact factor 4.2), *Current Medicinal Chemistry* (impact factor 3.8), etc. having more than 4,615 citations (H-Index ~31; i10 index:64; Cumulative Impact Factor ~330). He has filed 11 patent inventions and published 7 books with prestigious publishers, including Springer, Pan-Stanford, and Nova Publishers. Dr. Sharma has been felicitated with many awards for scientific excellence during his career: Outstanding Scientist Award (2021), Abdul Kalam Azad Award (2018), BRICPL Eminent Scientist Award (2017, 2018), Achiever Award (2017), Appreciation Award (2008, from UIC Illinois at Chicago, USA), MGIMS Young Scientist Award (2000), and VCS Memorial Young Scientist Award (1999).

Dr. Girish Kumar Gupta is presently working as Professor of Pharmaceutical Chemistry at Sri Sai College of Pharmacy, Badhani, and Director Research and Development at Sri Sai Group of Institutes, Pathankot, India. He has been awarded several fellowships and distinctions and has published more than 90 scientific publications. His current focus is on the discovery of bioactive natural products and chemical entities especially in the field of antimicrobials by the use of virtual screening followed by in vitro assays.

Dr. Mukesh Yadav is Assistant Professor at the Department of Biotechnology, Maharishi Markandeshwar (MM) (Deemed to be University), Mullana-Ambala, India. He accomplished his B.Sc. and M.Sc. in Biotechnology from Maharshi Dayanand University, Rohtak, India. He completed his Ph.D. degree in Microbial Biotechnology from Punjabi University, Patiala, India. He has been awarded with Indian fellowship programs (CSIR and DBT). His area of expertise is microbial biotechnology. He is extensively involved in research domain of fermentative production of microbial enzymes and microbial metabolites

https://doi.org/10.1515/9783110517736-203

of industrial importance. He has published more than 70 research and review articles in various reputed national and international journals. He has published books on microbial enzymes for food industries, industrial biotechnology, and therapeutic potential of flavonoids and instrumentation. He has life membership of several professional scientific societies and contributing to the development of society.

List of contributing authors

Chapter 1
Pooja Mittal
Maharishi Markandeshwar University
Sadopur, Ambala 134003
Haryana
India

Ramit Kapoor
Parexel International
Sadopur, Ambala 134003
Haryana
India

Rupesh K. Gautam
Maharishi Markandeshwar University
Sadopur, Ambala 134003
Haryana
India
Email id: drrupeshgautam@gmail.com

Chapter 2
Namrata Malik
School of Life Sciences
Jaipur National University
Jaipur 302025
Rajasthan
India
and
Institute of Agricultural Sciences
Banaras Hindu University
Varanasi 221005
Uttar Pradesh
India

Chapter 3
Malay Kumar Sannigrahi
Postgraduate Institute of Medical Education
and Research
Chandigarh
India
malaysannigrahi@gmail.com

Deepika
Thapar Institute of Engineering and
Technology
Patiala, Punjab
India
deepika0184@gmail.com

Chapter 4
Younis Ahmad Hajam
Department of Zoology
Career Point University
Hamirpur 176041
Himachal Pradesh
India

Rahul Datta
Centre for Agricultural Research and
Innovation
Guru Nanak Dev University
Amritsar 143005
Punjab
India

Sonika
Department of Zoology
Career Point University
Hamirpur 176041
Himachal Pradesh
India

Ajay Sharma
Department of Chemistry
Career Point University
Hamirpur 176041
Himachal Pradesh
India

Rajesh Kumar
Department of Zoology
Career Point University
Hamirpur 176041
Himachal Pradesh
India

https://doi.org/10.1515/9783110517736-204

Abhinay Thakur
Department of Zoology
DAV College
Jalandhar 144008, Punjab
India
Email Id: abhinay.thakur61@gmail.com

Anil Kumar Sharma
Department of Biotechnology
MMDU
Mullana, Ambala
Haryana
India
anibiotech18@gmail.com

Chapter 5
Amit Kumar Singh
Immunology and Microbial Diseases
Laboratory
Albany Medical Center
Albany, New York
NY 12208
USA
singha5@amc.edu

Chapter 6
Saloni Singh
School of Life Sciences
Jaipur National University
Jaipur 302025
Rajasthan
India
singhsaloni1990@gmail.com

Chapter 7
Anil Kumar
Department of Microbiology
DAV University
Jalandhar, Punjab
India
anilsharma2710@gmail.com

Shailja Sankhyan
Department of Biotechnology
Chandigarh University
Gharuan
Mohali, Punjab
India

Abhishek Walia
Department of Microbiology
CSK Himachal Pradesh Agricultural University
Palampur
Himachal Pradesh
India

Chayanika Putatunda
Department of Microbiology
DAV University
Jalandhar, Punjab
India

Dharambir Kashyap
Department of Histopathology
Postgraduate Institute of Medical Education
and Research
Chandigarh 160 012
Punjab
India

Ajay Sharma
Department of Chemistry
Career Point University
Tikker-Kharwarian
Hamirpur 176 041
Himachal Pradesh
India

Anil K Sharma
Department of Biotechnology
Maharishi Markandeshwar (Deemed to be
University) Mullana
Ambala, Haryana
India
anibiotech18@gmail.com

Chapter 8
Nazia Tarannum
Department of Chemistry
Chaudhary Charan Singh University
Meerut 250004
Uttar Pradesh
India
naz1012@gmail.com

Ranjit Hawaldar
Centre for Materials for Electronics
Technology
Pune, Maharashtra
India

Chapter 9
Preeti Kumari Sharma
Department of Microbiology
Government Medical College and Hospital
Jammu
India
and
Department of Biotechnology
Maharishi Markandeshwar (Deemed to be
University)
Mullana, Ambala
Haryana
India
priitisharma.micro@gmail.com

Paavan Singhal
Department of Biotechnology
Maharishi Markandeshwar
(Deemed to be University)
Mullana, Ambala
Haryana
India

Pooja Mittal, Ramit Kapoor and Rupesh K. Gautam

Chapter 1
Introduction of microbiology

Abstract: Microbiology is a very broad area covering all aspects related to microorganisms including bacteria, archaea, fungi, protozoa, algae, virus and multicellular animal parasites. Microbiology represents a diverse and complex array of research in medical sciences, food technology and pharmaceutical industry, and has a great impact on world's economy. This chapter deals with the basic aspects of microbiology, including microbial classification, cellular structure of bacteria and morphological forms of bacteria. The chapter also covers the basic aspects of microbial culture for the beginners.

Keywords: microbes; prokaryotes; eukaryotes; classification; culturing; diagnosis

1.1 Introduction

Bacteria were initially observed by Anton von Leeuwenhoek in the late seventeenth century but his research was not recognized until the initial stages of nineteenth century. In the golden age of microbiology, Robert Koch, Louis Pasteur and their associates contributed a lot to the cause of microbiology [1–3]. Their methods of culturing microbes involved the cultivation of the bacteria from the natural environment and growing them in the artificial environment according to the need.

The word microbes or germs have been generally classified into three categories, i.e., animal, mineral or vegetable. Microbes, also called microorganisms, are the minute structures which are not visible to the naked eye, such as bacteria, fungi, protozoa, algae and virus.

Usually, we associate the microbes only with infections, diseases, spoilage of food and their other destroying effects; however, they also have many important contributions in maintenance of balance in nature. Marine and other freshwater microbes are essential in maintenance of aquatic food chain while soil microbes are responsible for the decomposition of the waste products and maintenance of the nitrogen cycle between soil and environment. Many commercial applications are also attributed to microbes as they can be used for the production of wine, alcohols,

Pooja Mittal, School of Pharmaceutical Sciences, RIMT University, Mandi Gobindgarh, Punjab
Ramit Kapoor, Parexal International, Mohali
Rupesh K. Gautam, Maharishi Markandeshwar University, Sadopur, Ambala, Haryana 134003, e-mail: drrupeshgautam@gmail.com

https://doi.org/10.1515/9783110517736-001

organic products and acids. Neutraceuticals and other food industries also use microbes for the production of vinegar, pickles, ketchup, soy sauce, green olives, buttermilk, cheese, yoghurt and bread [3].

1.2 Microbial classification

To understand about the pathogenicity of the microbes, we must have the knowledge about classification of the microbes. First, we will have a look at the major groups of microbes and their pattern of classification.

The major groups are:
i.   Bacteria
ii. Archaea
iii. Fungi
iv. Protozoa
v.   Algae
vi. Virus
vii. Multicellular animal parasites

1.2.1 Bacteria

Bacteria (singular: bacterium) are simple unicellular microorganisms which contained prokaryotic (Greek meaning: prenucleus) cells as their genetic material is not enclosed by any membrane [2].

Bacterial cells may appear in various shapes like *bacillus* (rod like), *coccus* (spherical) and *spiral* (curved) mostly, while star-shaped and square-shaped bacteria were also seen some times. They may form some groupings, chains or clusters which are the characteristics of that particular genus or species. Bacterial cell wall generally consists of peptidoglycan (a carbohydrate and protein complex). They reproduce through the process known as binary fission which divides the bacterial cell into two equal parts. They feed themselves on the organic acids which are offered by the living or dead organisms of the nature. Bacteria can swim by using their appendages called flagella.

1.2.2 Archaea

Just like bacteria, they are also prokaryotes but their cell wall, if present, does not contain peptidoglycan. They exist in extreme environments. Based on their existence, they are categorized into three classes, *viz. methanogens* which release methane as their waste product, the *extreme halophiles* which live in the extreme salt environments like the Green Salt Lake and the Dead Sea and the *extreme thermophiles* thrive in the hot sulfurous water [3].

1.2.3 Fungi

These eukaryotes possess distinct nuclear matter which is covered by nuclear envelop or nuclear membrane. They can be unicellular or multicellular. Mushrooms are the typical example of multicellular fungi which have appearance like plants but photosynthesis process is not present in them. Chitin is present inside the cell wall of true fungi. The unicellular yeast has been known to be a primitive eukaryote which is bigger in size than bacteria.

1.2.4 Protozoa (singular – *protozoan)*

They are unicellular eukaryotes that move with the help of pseudopods (false feet), cilia or flagella. They contain variety of shapes and can live as free entities or as dependent on host (parasites). They can have both sexual and asexual modes of reproduction [4].

1.2.5 Algae

They are the eukaryotes that have photosynthetic ability. They can reproduce by sexual as well as asexual reproduction. The cell wall of algae is mostly composed of a complex carbohydrate called cellulose.

1.2.6 Virus

They are the different species which are acellular in nature. They contain single type of genetic material, either DNA (deoxyribonucleic acid) or RNA (ribonucleic acid) in the core, which is covered by a protein coat. They are dependent on other microbes for their reproduction. Thus, infection to the host cell is the proof of their livingness [5].

1.2.7 Multicellular animal parasites

They are not strictly the microorganisms but they are of intense medical importance. They thrive on the host animals. Majorly, two types of parasites exist, i.e., round worms and flat worms also called as helminthes [2, 4–6].

1.3 Bacterial morphology and cytology

Basically, the cells are divided into two basic types, i.e., prokaryotic cells and eukaryotic cells. Eukaryotic cells are more complex than prokaryotes. Plants and animal cells are of eukaryotic types, while bacterial cells are prokaryotic in nature. Bacteria and archaea are categorized into prokaryotes, while fungi, protozoans and algae are eukaryotes by nature. On the other hand, viruses are noncellular structures that do not fit their self in any of the above classifications [7, 8].

The bacterial cell generally ranges from 0.2 to 2 µm in diameter and 2 to 8 µm in length. They possess the basic shapes like spherical **coccus**, rod-shaped bacteria or **bacillus** and spiral shaped. **Coccus** bacteria are generally round in shape but they can also be elongated, oval, flattened or one-side flattened. Cocci reproduce by cell division and sometimes they remain attached to each other called as **diplococci**, while **streptococci** exist in the form of chains. The microbes that divide themselves into two equal planes and present themselves in the pattern of four are called **tetrads**, while those divided in three planes and present themselves in a cubic structure of eight are known as **sarcinae**. While **staphylococci** exist like bunches of grapes after dividing into multiple equal planes.

Bacilli are rod-shaped bacteria so they divide themselves along their longitudinal axis only. Bacteria that divide and appear themselves in pairs are called **diplobacilli** while those that appear in chain-like structures are called as **streptobacilli**. Some bacilli are oval in shape and appear like cocci and are called **coccobacilli** [1, 2, 4, 7, 8].

Spiral bacteria have one or more twists. Those that appear in curvy rod shape are called vibrios, while others that appear in helical shapes like cork screw with rigid structures are called spirilla. The helical-shaped bacilli which have a flexible body are called **spirochetes**. The shape of bacteria is generally influenced by the heredity. Most of the bacteria are **monomorphic** genetically which means that they remain in same shape during all conditions, while some bacteria are **pleomorphic** which means that they can have more than one shape, for example, *Rhizobium* and *Corynebacterium*.

1.3.1 Structural components of bacterial cell/prokaryotic cell

The structures exterior to the prokaryotic cell wall is glycocalyx, pilli, fimbriae and flagella.

1.3.1.1 Glycocalyx

Glycocalyx is the substance that is secreted by the prokaryotic cell onto their cell wall. Bacterial glycocalyx is viscous and sticky substance which is a polysaccharide or protein or both chemically. It is also known as **capsule** when it is firmly attached to the cell wall but if it is loosely attached and unorganized, then it is called **slime layer**. Sometimes, capsules protect the pathogenic bacteria from the phagocytosis process in the host cells, for example, *Bacillus anthracis* produces a capsule of D-glutamic acid because of which it causes anthrax and gets escaped from the phago-cytosis caused by the host cell [1, 2, 9].

1.3.1.2 Flagella

Flagella are the filamentous appendages present on the exterior surface of some bacteria which help in the movement of the bacterial cells. Cells without flagella are known as atrichous (without projections). Flagellum consists of three basic parts, i.e., filament, hook and basal body. Filament is the long outermost region which is constant in diameter and contains a globular protein flagellin. Filament is attached to the bacterial cell with the help of hook which consists of different proteins, while basal body anchors the flagellum to the cell wall [6, 10–12].

1.3.1.3 Cell wall

The bacterial cell wall is composed of peptidoglycan or murein which can be present alone or in the combination of others. The polysaccharides are made up of repeating units of disaccharides which are attached by polypeptides and form a lattice structure which will protect the whole cell and its components. The disaccharide portion of the cell wall is made up of two monosaccharides, namely *N*-acetyl glucosamine and *N*-acetyl muramic acid. Cell wall of **gram-positive bacteria** consists of many repeating layers of peptidoglycan which form a stiff structure. Also, the gram-positive bacterial cell wall contains teichoic acid which contains an alcohol and a phosphate. Teichoic acids are negatively charged molecules which bind to cations inside the cell wall and regulate their movement inside or outside the cell wall. In contrast, the gram-negative bacterial cell wall consists of fewer layers of peptidoglycan which is bounded to the lipoproteins present in the outer layers of cell membrane. Also, techoic acids are not present inside the cell wall of these bacteria. The cell wall of gram-negative bacterial cell is much susceptible to mechanical breakage because of the lack of the thick coat of peptidoglycan [9, 13–15].

Structures interior to the cell wall: Inside the bacterial cell, cytoplasm, cytoplasmic membrane, nucleoid, inclusion bodies and ribosomes are present.

1.3.1.4 The cytoplasmic/plasma membrane

This is a phospholipid and protein containing thin membrane which encloses the cytoplasm of the cell. It was observed under electron microscope that this membrane consists of two layers with a space present between the layers. It consists of a phospholipid bilayer structure with hydrophilic heads at the exterior parts while hydrophobic tails in the interior part. The polar head consists of a phosphate group and a glycerol molecule which are responsible for the hydrophilicity, while the tails are composed of fatty acids that render them hydrophobic by nature. The protein can be arranged in a variety of ways onto the lipid bilayer cell membrane. They can be peripherally attached or integrally attached to the cell membrane. The proteins that are attached peripherally are easy to remove by mild treatments while those that are integrally attached require tough procedures for elimination. Some carbohydrates are attached to the outer surface of the cell membrane which are called as **glycoproteins**, while the lipids that are attached to the glycoproteins are called **glycolipids**. This model depicting dynamic arrangements of lipids and protein onto the outer layer of lipid bilayer cell membrane is called **a fluid mosaic model** [10, 14].

1.3.1.5 Cytoplasm

The liquid viscous substance present inside the cell membrane is called cytoplasm. Composition of cytoplasm is about 80% water, proteins, inorganic ions, carbohydrates, lipids and low-molecular-weight substances. It is thick, semitransparent, aqueous and elastic in nature.

1.3.1.6 The nucleoid

The nucleoid of the cell contains the genetic information incorporated inside in the form of DNA and RNA. Double-stranded DNA or the bacterial chromosome is arranged as single thread circulated inside the nucleoid. Bacterial chromosomes do not contain histones and any nuclear membrane.

In addition to the bacterial chromosomes, cells also contain the double-stranded DNA molecules in the form of small circular vesicles which are called as **plasmids**. They are extrachromosomal material, i.e., they are independent of the bacterial chromosomes and replicate independently [2, 12].

1.3.1.7 Ribosomes

The process of protein synthesis occurs at the sites of ribosomes. The cells that synthesize proteins at a higher rate possess more ribosomes as compared to the others. They consist of two subunits, i.e., smaller and larger subunits each of which contains the proteins and type of ribosomal RNA. Prokaryotic ribosomes are 70S type and have 50S as larger subunit and 30S as smaller subunit. The letter S is the Svedberg unit which indicates their relative sedimentation rates during ultrahigh centrifugation procedures [16–18].

1.3.1.8 Inclusions

In the cytoplasm, some reserves are there which are known as inclusion bodies. These reserves contain some nutrient molecules which are to be used by the cell during the harsh conditions. Some metachromatic granules, iron granules and polysaccharides granules are the typical examples of the reserve deposits inside the inclusions.

1.4 Bacterial/microbial culture

A bacterial culture is a process of multiplication of the microbes by providing them with conditions required for their reproduction under controlled laboratory conditions. These methods are the basic and fundamental tools of the research, for example, to evaluate the antimicrobial properties of any drug product or the formulation, we need to prepare the microbial culture. It is one of the primary diagnostic methods of microbiology and used as a tool to determine the cause of infectious disease by letting the agent multiply in a predetermined medium.

It is necessary to separate a pure culture of microorganisms. A pure (or *axenic*) culture is defined as a population of microbial cells that grow in the absence of other species or other microbes. It can be obtained from a single cell.

There are several types of culture methods which are named according to the name of the material being cultured.

1.4.1 Broth cultures

This is a liquid culturing method in which the desired bacteria are suspended in the liquid nutrient medium such as Luria broth. This method allows the scientist to grow large amount of bacteria. This type of culturing is of utmost importance where the microbial inoculum is cultured overnight in the appropriate medium, viz. Luria Bertani broth and nutrient broth, and then culture samples are taken for further testing of antimicrobial activity of the drug.

1.4.2 Agar plate culture

Thin layer of agar-based nutrient medium is spread over the Petri plates of different sizes which are inoculated with desired bacteria and kept at optimal conditions for the multiplication of the bacteria. Once the desired population of the microbes is achieved, the plates are removed and stored in upright position in the refrigerators for further use.

1.4.3 Stab culture

This is similar to the agar culture but the difference lies in the fact that solid agar media is used in test tubes for the preparation of stab culture. Here, bacteria are inoculated via inoculating needle or by pipette in the center of the stab. They are basically used for short-term storage [19, 20].

After these introductory words, through this book you will have the understanding about the principles of medical microbiology where we will study about various types of microbes, including bacteria, fungi and viruses, along with their underlying significance. This book will thoroughly introduce you to the colossal variety of microbes, their classification and the various techniques and procedures which the microbiologists employ to study and visualize the microbes.

References

[1]   Hugo WB, Russell AD, Pharmaceutical Microbiology. Blackwell Science, 1998.
[2]   Tortora GJ, Funke BR, Case CL, Johnson TR, Microbiology: An Introduction, Benjamin Cummings San Francisco, CA, 2004.
[3]   Willey JM, Sherwood L, Woolverton CJ, Prescott, Harley, and Klein's Microbiology, McGraw-Hill Higher Education, 2008.
[4]   Baron S, Classification–Medical Microbiology, University of Texas Medical Branch at Galveston, 1996.
[5]   Eisen JA, Assessing evolutionary relationships among microbes from whole-genome analysis, Curr Opin Microbiol, 2000, 3(5), 475–80.
[6]   Raoult D, How the virophage compels the need to readdress the classification of microbes, Virology, 2015, 477, 119–24.
[7]   Shimazu K, Takahashi Y, Uchikawa Y, Shimazu Y, Yajima A, Takashima E. et al., Identification of the Streptococcus gordonii glmM gene encoding phosphoglucosamine mutase and its role

in bacterial cell morphology, biofilm formation, and sensitivity to antibiotics, FEMS Immunol Med Microbiol, 2008, 53(2), 166–77.

[8]　Sousa AM, Machado I, Nicolau A, Pereira MO, Improvements on colony morphology identification towards bacterial profiling, J Microbiol Methods, 2013, 95(3), 327–35.

[9]　Knaysi G, Elements of bacterial cytology, AJN the Am J Nurs, 1945, 45(2), 166.

[10]　Bisset K, Bacterial cytology, Intl Rev Cytol Elsevier, 1952, 93–106.

[11]　Knaysi G, Cytology of bacteria, Bot. Rev., 1938, 4(2), 83–112.

[12]　Marshak A, Bacterial cytology, Intl Rev Cytol Elsevier, 1955, 103–14.

[13]　Ghuysen J-M, Use of bacteriolytic enzymes in determination of wall structure and their role in cell metabolism, Bacteriol Rev, 1968, 32(4 Pt 2), 425.

[14]　Kulczycki E, Ferris F, Fortin D, Impact of cell wall structure on the behavior of bacterial cells as sorbents of cadmium and lead, Geomicrobiol J, 2002, 19(6), 553–65.

[15]　Woo I-S, Rhee I-K, Park H-D, Differential damage in bacterial cells by microwave radiation on the basis of cell wall structure, Appl Environ Microbiol, 2000, 66(5), 2243–47.

[16]　Shajani Z, Sykes MT, Williamson JR, Assembly of bacterial ribosomes, Annu Rev Biochem, 2011, 80, 501–26.

[17]　Shine J, Dalgarno L, Determinant of cistron specificity in bacterial ribosomes, Nature, 1975, 254(5495), 34.

[18]　Wittmann H, Components of bacterial ribosomes, Annu Rev Biochem, 1982, 51(1), 155–83.

[19]　Orphan V, Taylor L, Hafenbradl D, Delong E, Culture-dependent and culture-independent characterization of microbial assemblages associated with high-temperature petroleum reservoirs, Appl Environ Microbiol, 2000, 66(2), 700–11.

[20]　Zambrano MM, Siegele DA, Almiron M, Tormo A, Kolter R, Microbial competition: Escherichia coli mutants that take over stationary phase cultures, Science, 1993, 259(5102), 1757–60.

Namrata Malik
Chapter 2
Bacteriology

Abstract: Bacteria are unicellular and prokaryotic microbes of immense economic importance. Bacteria may be pathogenic, toxin producing, or beneficial for human and other lives. Bacteria show a wide range of variations in morphology, culture, genetics, immunological responses and other biological aspects. This chapter covers all major aspects related to bacteria. In this chapter, bacterial morphology, culture aspects, immunological aspects, bacterial genetics have been discussed. Also, the bacterial growth curve with various life phases has been discussed. Major bacterial diseases have been added from medical microbiology point of view.

Keywords: bacterial morphology, metabolism, pathogenesis, virulence, immune responses, bacterial diseases

2.1 Bacterial morphology and structures: metabolism and growth

2.1.1 Bacterial morphology

Bacteria based on their cellular organization and biochemical properties are classified under the kingdom Protista. This kingdom is further categorized into prokaryotes and eukaryotes. Bacteria and blue green alga are grouped under prokaryotes, whereas fungi, algae, slime molds and protozoa are eukaryotes. Major characteristics of bacteria are based on shape, size and arrangements. Morphology is the systematic study of external features of bacteria. The basic bacterial morphologies are cocci (spherical in shape), bacilli (rod shaped) and other various shapes such as spiral, vibrio and helical.

Morphology deals with the study of size, shape and arrangement of bacteria [54].
i.   **Size**: bacterial size is measured by micrometry. The limit of resolution, i.e., the minimum distance between two points at which they can be visually distinguished as separate with the naked eye, is about 200 μm. Due to their smaller size, bacteria can be observed only with the aid of microscopes. Bacteria of medical importance generally measure about 0.2–1.5 μm in diameter and about 3–5 μm in length.

Namrata Malik, School of Life Sciences, Jaipur National University, Jaipur 302025, Rajasthan, India; Department of Botany, Institute of Science, Banaras Hindu University, Varanasi 221005, India, e-mail: namratamlk@gmail.com

https://doi.org/10.1515/9783110517736-002

ii. **Shape**: the basic bacterial shapes are coccus (spherical in shape), bacillus (rod shaped) and spiral which are further classified as spirilla, spirochetes, or vibrio (comma shaped) [23].

a. **Coccus**: The cocci (plural) are the bacteria in shape of tiny balls. They can be arranged in the following ways based on their plane of division:
   - Division in one plane produces bacteria arranged in pairs (also known as **Diplococcus**, e.g., *Neisseria gonorrhoeae* and *N. meningitidis*) or in chains (e.g., *Streptococcus pyogenes*)
   - Division in two planes produces **Tetrad**, four cells forming a flat square (e.g., *Micrococcus roseus*)
   - Division in three planes produces **Sarcina**, cube like group of eight cocci (e.g. *Sarcina maxima* and *Sarcina ventriculi*)
   - Division in random planes produces irregularly arranged cells (e.g. *Staphylococcus aureus*)

* An average coccus is about 0.5–1.0 µm in diameter [32].

b. **Bacillus**: The bacilli (plural) are rod shaped and appear longer than they are wide (e.g., *Bacillus anthracis*). They divide in one plane and can be arranged as streptobacilli:
   - **Streptobacilli**: bacilli are arranged in chains
   - **Coccobacilli**: they appear oval and alike coccus

* An average bacillus is 0.5–1.0 µm wide by 1.0–4.0 µm long.

c. **Spiral**: they appear twisted in helices and resemble cork screws. They can be a vibrio, a spirillum and a spirochete:
   - **Vibrio**: comma shaped (e.g., *Vibrio cholerae*)
   - **Spirillum**: it is thick and rigid spiral (e.g., *Helicobacter pylori*)
   - **Spirochete**: it is thin and flexible spiral (e.g., *Treponema pallidum*)

* Spirals range in size from 1 µm to over 100 µm in length.

2.1.1.1 Why the bacteria live with different shapes?

It is proved that the rod-shaped organisms raised first and the coccoid being the last. It is possible that various morphologies of bacteria might contribute to natural selection [60]. The survival of organisms through the evolutionary forces can be attributed to the fulfillment of the following selection criterion:
- Nutrient uptake
- Cell division
- Predation

– Attachment to surfaces
– Active motility

The above factors are important because they all will decide whether the cells are going to survive or not as they must grow and multiply to be alive.

2.1.2 Bacterial growth

Growth occurs when bacteria divide by binary fission forming two daughter cells. It results when there is an increase in cell size or number. **Coenocytic** microorganisms, i.e., a multinucleate cell, are formed when nuclear division is not followed by cytoplasmic division. The factors required for bacterial growth are discussed here:
i.   Environmental factors
ii. Sources of metabolic energy

2.1.2.1 Environmental factors affecting growth

a. **Nutrients:** bacterial cell requires macroelements/macronutrients and microelements/micronutrients for its growth (Table 2.1). The microelements do not limit the growth of bacteria.

Table 2.1: List of nutrients with their role in bacterial metabolism.

Type of nutrient	Functions/uses
Macronutrients	
Carbon, oxygen, hydrogen, nitrogen, sulfur, phosphorus	Components of carbohydrates, lipids, protein and nucleic acid
Potassium (K^+)	Required by enzymes
Calcium (Ca^{2+})	Heat resistance of bacterial spores
Magnesium (Mg^{2+})	Cofactor for many enzymes, complexes with ATP, stabilizes ribosomes and cell membrane
Iron (Fe^{2+} and Fe^{3+})	Part of cytochrome and a cofactor for enzymes and electron-carrying proteins
Micronutrients	
Manganese (Mn^{2+})	Helps in catalysis of phosphate group transferring enzymes
Molybdenum (Mo^{2+})	Nitrogen fixation
Cobalt (Co^{2+})	Component of vitamin B_{12}

Autotrophs: these types of microbes use carbon dioxide as their only source of carbon and can synthesize their own food by photosynthesis.

Heterotrophs: they use preformed organic molecules as carbon source.

Mixotrophic: this type is a combination of both autotrophs and heterotrophs. They depend on both inorganic and organic sources of carbon (e.g., *Beggiatoa*).

b. pH: it is the negative logarithm of hydrogen ion concentration and can be expressed as follows:

$$pH = -\log [H^+]$$

It is a major factor affecting bacterial growth, and every species has an optimum pH range for its growth and survival (Table 2.2).

Table 2.2: Bacterial characterization based on pH requirements.

Category	pH range	Example
Acidophiles	0.0–5.5 (acidic conditions)	Fungi
Neutrophiles	5.5–8.0	Most bacteria and protozoa
Alkalophiles	8.5–11.5 (basic conditions)	*Natronobacterium*
Extreme alkalophiles	pH 10 or higher	*Bacillus alcalophilus*

c. Gaseous requirement: Microbes growing in the presence or absence of oxygen are known as aerobic or anaerobic, respectively. They are further classified as follows:
- **Obligate aerobes:** they are completely dependent upon atmospheric oxygen for their growth and survival (e.g., fungi and algae)
- **Facultative anaerobe:** they can grow in the absence of oxygen, but if allowed to grow in presence of oxygen they witness a better growth (e.g., yeast, Enterobacteriacea)
- **Aerotolerant aerobes:** they are indifferent to the presence or absence of oxygen (e.g., *Enterococcus faecalis*)
- **Strict or obligate anaerobes:** they are killed in the presence of oxygen (e.g., *Bacteroides gingivalis* which lives in the anaerobic crevices of buccal cavity, and *Fusobacterium*)
- **Microaerophiles:** they require very low concentration of oxygen (2–10%) usually below the normal atmospheric level of oxygen (20%) (e.g., *Campylobacter*)

All the bacteria and protozoans show various modes of oxygen requirements. Generally, oxygen leads to inactivation of proteins and formation of toxic products. Oxygen is readily reduced (due to the presence of two unpaired electron in the outermost shell) and forms superoxide radical, hydrogen peroxide and hydrogen radical.

The products formed by reduction of oxygen are powerful oxidizing agents and have all the strength to destroy cellular components. To overcome the ill effects of oxygen reduction, obligate aerobes and facultative anaerobes usually contain the enzyme superoxide dismutase (SOD) and catalase. The SOD catalyzes the dismutation of the superoxide radical into ordinary molecular oxygen and hydrogen peroxide. The catalase catalyzes the decomposition of hydrogen peroxide to water and oxygen.

The anaerobic microbes lack both these enzymes and therefore cannot tolerate oxygen.

d. Temperature: The external surrounding temperature decides the cell's internal environment conditions. The enzymes that are required for cell's metabolism are dependent on temperature for its functioning. If the cell's temperature goes too high or low, its metabolism is affected. Table 2.3 describes the categories of bacteria based on temperature requirements.

Table 2.3: Categorization of microbes based on temperature requirements.

Type	Optimum temperature range	Habitat/example
Psychrophiles	0–15 °C	Arctic and Antarctic habitats, *Pseudomonas*, *Vibrio*, *Alcaligenes*, *Bacillus*
Mesophiles	20–30 °C	All human pathogens
Thermophiles	55 °C or higher	Compost, hot water lines, hot springs
Hyperthermophiles	80–100 °C	*Pyrodictium occultum*

2.1.3 Bacterial growth curve

The bacterial cells are grown and incubated in a liquid medium (broth) in a batch process. In a batch process, the cell growth occurs in a closed system and no fresh nutrient medium is added during incubation. Bacterial counts are done at intervals after inoculation and plotted in relation to time; a growth curve is obtained [50]. The curve has four distinct phases:

i. **Lag phase:** This is the first phase of the bacterial growth curve. The microorganisms are introduced to a fresh culture medium, and an increase in cell size is observed. The cell division process takes time to adapt to new environmental conditions. During this phase, microbes undergo vigorous metabolic activity but do not divide. The present media requires new enzymes for nutrients to be absorbed in fresh medium. Cells start replicating their genome and finally start dividing. This stage is affected very less due to antibiotics. Slowly bacteria get adjusted to the new medium. It must be noted that longer lag phase corresponds to the fact that microbes are transferred to a new media and shorter lag phase corresponds to the same media.

ii. **Logarithmic phase/trophophase/exponential phase**: During this phase, maximum utilization of media takes place, and a linear relationship between time and logarithm of the number of cells is established. It is a period of fast division. This type of growth is balanced, i.e., when cellular components are formed at constant rates. Unbalanced growth is generally observed when the cells are transferred from a nutrient-depleted medium to nutrient-rich medium or vice versa. Hence, in nutrient-rich medium, proteins are formed first, for which ribosomes are formed at a faster rate. Whereas in nutrient-deficient medium, enzymes need to be synthesized for metabolism of unavailable nutrients. The exponential growth takes place until one or more nutrients in the medium is exhausted, or toxic metabolites accumulate and inhibit growth. During this phase, the biomass increases exponentially with time. The average time required for the population, or the biomass to double is known as the generation time or doubling time. A linear plot is observed when logarithm of biomass concentration is plotted as a function of time. Antibiotics act better in this phase.

iii. **Stationary phase/idiophase**: This phase is marked by equilibrium between death rate and growth rate [33]. The death of cell occurs due to production of secondary metabolites, toxins or nutrient limitation.

iv. **Decline phase/death phase**: The death of cells occurs. The remaining living bacteria use the remnants of dead bacteria as nutrient and survive as persister.

2.1.4 Calculations of growth

The time taken by bacteria to double its number is called generation time or doubling time. It is represented by "g." For example, in a laboratory experiment, a conical flask containing 100 mL of Luria Bertani culture media is inoculated with one cell of *E. coli*. The cell divides by binary fission and produces a daughter cell in 20 min. These two cells further divide and produce four cells in another 20 min. The number of cells is doubling in every 20 min, which is said to be the generation time of *E. coli*.

There is logarithmic increase in the number of cells and is represented by 2^n, where n denotes the number of generations. Also, the number of generations produced per unit time is known as growth rate [8].

To solve for generation time g and growth rate k:

Let, N_o be the initial number of cells
N_t be the number of cells after time t
n be the number of generations after time t
Then, $N_t = N_o \times 2^n$
Solving for n and using logarithm to the base 10,
$\log N_t = \log N_o + n \log 2$
further $n = \log N_t - \log N_o/\log 2$
putting value of $\log 2 = 0.3010$
$n = \log N_t - \log N_o/0.3010$
the mean growth rate (k) is expressed as n/t, i.e., the number of generations per hour
$k = n/t = \log N_t - \log N_o/0.3010t$
if g is the time during which number of cell doubles its initial number, i.e.
$N_t = 2N_o$ then $t = g$
$k = \log(2N_o) - \log N_o/0.3010g$
$k = \log 2 + \log N_o - \log N_o/0.3010g$
$k = 1/g$

Thus, the generation time and mean growth rate constant are inversely proportional to each other.

2.2 Bacterial heredity and variation

2.2.1 Heredity and variation

Heredity refers to the similarity between parents and progeny, whereas variation is the difference among them. Variations can be due to genetic changes (heritable) or due to environmental changes (nonheritable). The study of heredity and variation is genetics. Gene, a unit of heredity, can be defined as a segment of DNA specifying for a particular polypeptide. The noncoding sequences of a gene are known as introns, whereas exons are its coding part. Genes can be identified based on their contribution to phenotype. Changes at genotypic level bring variations in the phenotype. Bacterial genetics is used as a model to understand DNA replication, genetic characters, their changes, and transfer to next generations.

2.2.1.1 Genetic information in bacteria

2.2.1.1.1 Chromosome
The prokaryotic microbes have the entire genome packed into a single, irregularly folded compact mass called nucleoid or bacterial chromosome [59]. DNA is double

helical and circular in structure. The bacterial chromosome is in direct contact with the cytoplasm, since a nuclear membrane is absent. It shows bidirectional replication, i.e., replication starts at one point but moves in both directions. The mode of replication is always semiconservative [42]. It begins at the origin of replication (*ori C* in *E. coli.*) and terminates at *ter* sites. The two daughter chromosomes are separated before cell division so that each cell gets one of the replicated DNA. The bacterial cells are haploid as they contain a single copy of chromosome.

2.2.1.1.2 Plasmid

They are the extrachromosomal DNA or genetic material in the cytoplasm. Plasmids when incorporated with bacterial chromosome are known as **episome**. Plasmids are capable of replicating independently [52] and control most of bacterial cell functions such as antibiotic resistance controlled by R plasmid. Plasmids can be lost or transferred from one bacterium to another. They can also be classified by incompatibility typing. The unrelated plasmids can easily live in the same bacterial cell but similar ones cannot coexist.

2.2.1.1.3 Transposons

They are also known as jumping genes. Transposons are not capable of carrying out their own replication but are dependent upon their integration with bacterial DNA [6].

2.2.1.1.4 Phage

Phages are extremely diverse in their genetic contents and their mode of replication. There are two types of phages: lytic and temperate. Lytic phages produce many copies of themselves in a single cell and release them after cell lysis, whereas temperate phage gets integrated with bacterial genome known as **prophage** and get replicated along with bacterial genome.

2.2.2 Genotypic and phenotypic variation

Genotype includes the complete genetic constituent of the cell, whereas the phenotype is the physical expression of genotype in a given environment. For example, typhoid causing *Salmonella typhi* is normally flagellated but if grown in phenol agar, the flagella are not synthesized. Another example is beta galactosidase production by *E. coli*. The gene becomes active only in the presence of lactose. If only glucose is present in the media, then the enzyme is not synthesized. Such substrate-induced enzymes are called as induced enzymes, whereas constitutive enzymes are the ones that are constantly expressed. The phenotypic variations are temporary, nonheritable and influenced by environment. Genotypic variations are stable, heritable and uninfluenced by the environment. The genotypic variations are brought about by mutation or any of the three gene transfer mechanisms [13], which are transformation, transduction and conjugation.

2.2.2.1 Mutation

Mutation is caused by alteration of the genetic sequences and can be due to addition, deletion or substitution of one or two nucleotide pairs in the coding regions of a gene. Mutations arise either spontaneously, i.e., arise occasionally on their own without any external agent known as **spontaneous mutation**, whereas mutation generated due to any physical or chemical agent (mutagen) is known as **induced mutation**.

2.2.2.1.1 Spontaneous mutation

They mainly occur due to transposons or errors in genetic replication. Errors in replication can be due to tautomerism, i.e., a template nucleotide takes on a tautomeric form. Tautomers are basically different forms of the nitrogenous bases. The bases normally exist in keto form but takes imino or enol form due to tautomeric shifts (change in the hydrogen bonding characteristics of the bases). Due to tautomerism, purine for purine or pyrimidine for pyrimidine substitutions are possible. These substitutions are known as **transition mutations**. If a purine gets substituted by pyrimidine or *vice versa*, then it is known as **transversion mutations**, though transverse mutations are bit rare because of steric problems between the bases. Replication errors also lead to various other kinds of alteration in nucleotides.

2.2.2.1.2 Induced mutation

As stated earlier, mutagens causing induced mutation are classified according to their mode of action. Mutagens are physical, chemical or biological agents that damage the DNA, which result in faulty replication. The chemical mutagens can be base analogs, DNA-modifying agents such as alkylating agents, deaminating agents and flat aromatic compounds acting as intercalating agents. Radiations are the physical mutagen damaging the DNA. **Base analogs** are structurally analogous to nitrogenous bases and can be incorporated into DNA during replication. Examples of base analogs are 5-bromouracil and 5-aminopurine. 5-Bromouracil is an analog of thymine, which tautomerize more readily than normal base. The enol tautomer of thymine now behaves as cytosine which pairs with guanine, thus leading to an unusual thymine and guanine pairing. **DNA-modifying agents** change the base structure and alter their base pairing preferences, for example, ethyl methanesulfonate (alkylating agents), hydroxylamine (hydroxylate cytosine) and nitrous oxides (NO, deaminating agent). Cytosine is hydroxylated at C-4 nitrogen which now base pairs with thymine. **Intercalating agents** such as proflavin and acridine orange stack themselves between the hydrogen bonds among the nitrogenous bases resulting in mutation by distortion of DNA helix.

2.2.2.2 Transmission of genetic material

2.2.2.2.1 Transformation

It is the direct uptake of donor DNA by recipient cells. Griffith in 1928 found that mice could not survive when injected with a mixture of live noncapsulated (rough, R) *Streptococcus pneumoniae* (nonvirulent) and heat-killed capsulated (smooth, S) strains of *S. pneumoniae*. These live R and heat-killed S strains were not fatal when used separately. Upon autopsy, these experimented mice were found to contain live virulent cells of *Streptococcus pneumoniae*. These experiments and further experiments established that these cells were recombinant of heat-killed virulent and live avirulent cells. It was concluded that a genetic exchange occurred between the dead cells and live ones. Such transformation was demonstrated in vitro also, and the transforming element was identified as DNA by Avery, MacLeod and Mc Carty in 1944 [3]. Transformation is possible only in the presence of competence factor produced during growth cycle [57]. Naturally competent transformable bacteria are found in several genera and include *Bacillus subtilis, Haemophilus influenzae, Streptococcus pneumoniae* and *Neisseria gonorrhoeae*.

2.2.2.2.2 Transduction

The phage-mediated transfer of DNA from one bacterium to another is known as transduction [40]. Bacteriophages consist of protein capsid and a nucleic acid core similar to any normal phage. They have the capability to introduce bacterial DNA derived from the cell in which it is developed into another sensitive cell, resulting in transfer of genetic material between two cells. The recipient cell acquires new characteristics coded by donor DNA. Transduction can be generalized when any segment of donor DNA is involved, or it may be specialized when only a particular segment of DNA is transduced. Episomes and plasmid may also be transduced.

2.2.2.2.3 Conjugation

J. Lederberg and E. L. Tatum in 1946 discovered genetic exchange between male or donor bacterium with a female or recipient bacterium *via* physical contact [36]. This is equivalent to sexual reproduction in higher organisms. Conjugation was first described in a strain of *E. coli* called K12. The donor or male cells designated as F+ (F plasmids are present which codes for sex pilus or conjugation tube which erupts through cell's surface through which the physical contact occurs) and F− cells (lacking F plasmid) act as females. F plasmid carries all the genes encoding essentials for conjugative gene transfer, i.e., genes for conjugation tube synthesis but lacks any identifiable genetic marker such as drug resistance [28].

2.3 Bacterial infection and pathogenesis

2.3.1 Pathogenesis

Pathogenesis is the study of bacterial infection including the initiation of the infectious process and the mechanisms leading to the development of signs and symptoms of disease. Pathogenicity is the ability to produce disease in a host organism. Characteristics of pathogenic bacteria are transmissibility, adherence to host cells, persistence, invasion of host cells and tissues, toxigenicity and the ability to evade or survive the host's immune system [14]. Pathogenicity is expressed by means of virulence, referring to the degree of pathogenicity of the microbe. Virulence comes with resistance of bacteria to antimicrobials and disinfectants [30]. Most of the infections are asymptomatic, and symptoms are seen when immunologic reactions are generated after infection.

2.3.1.1 Types of bacterial pathogens

Bacterial pathogens can be classified into two groups: primary and opportunistic pathogens. Those pathogens that can establish infection in a healthy individual are termed as primary pathogen, whereas opportunistic pathogens more readily cause disease in individuals with weakened immune system. Many opportunistic pathogens, e.g., coagulase-negative staphylococci and *Escherichia coli*, are present normally in humans but introduction of these bacteria onto sites where they are not normally found can cause development of the disease. Many of the *Neisseria meningitidis* are harmless strains but are considered opportunistic pathogens. Also, *Mycobacterium tuberculosis* is a primary pathogen but does not cause infection in each individual it invades.

2.3.1.2 Koch's postulates

Koch's postulates establish a relationship between pathogen and disease. He gave four criteria to establish the relationship. The postulates were jointly formulated by Robert Koch and Friedrich Loeffler in 1884 and refined and published by Koch in

1890. Koch studied the disease anthrax caused by *Bacillus anthracis* to design the postulates but they are now generalized to other diseases.

a. The organism must always be found in infected humans but they should not be present in healthy individuals.
b. The organism must be isolated from the diseased individual and can be grown in pure culture.
c. The organism isolated in pure culture must initiate disease when reinoculated into susceptible animals.
d. The organism must be reisolated from the experimentally infected animal.

2.3.1.3 Transmission of infection

The bacteria producing diseases without any prominent symptoms are transferred from one person to another. Some of the pathogens normally reside in animals and incidentally affect humans. The anthrax causing bacteria *Bacillus anthracis* normally present in the environment affect animals and occasionally infect humans through contaminated animal products. *Salmonella* and *Campylobacter* species typically infect animals and transmitted *via* food to animals. Another species of *Clostridium* is present in environment and transmitted to humans by ingestion such as gastroenteritis causing *C. perfringens* and botulism causing *C. botulinum*. If wound comes in contact with soil, then gas gangrene causing *C. perfringens* and tetanus causing *C. tetani* attacks. Diseases are often spread and promoted by clinical manifestations such as sputum, cough and genital discharge due to diseases. *Staphylococcus aureus* if present on patient's hand is transferred to another healthy one by touching. **Nosocomial infections** are brought about when a person gets infected *via* the hospital personnel.

Entry of pathogens

The pathogen entry in the host cell can be through:
a) Respiratory tract (both upper and lower)
b) Gastrointestinal tract
c) Genital pathways
d) Urinary tract
e) Cuts, burns and other injuries

The pathogen must cross these barriers for generating disease. Every infectious disease has a characteristic sign and symptom.

Source of infection: Few of the common sources of infection are humans, animals, insects, soil, water and food.

Humans: Pathogen is present in a patient or carrier. A **carrier** is a person or animal with asymptomatic infection and that can be transferred to another susceptible person or animal. There are several types of carrier: *healthy* carrier is the one who is carrying the pathogen but has never developed the disease. A *covalescent* carrier is the type of carrier which has recovered from the disease but still carries the pathogen. Carriers can also be temporary and chronic. *Temporary* carriers are the ones which carry the pathogen for shorter duration of time, i.e., less than 6 months, whereas the time period for carrying the pathogen in case of *chronic* carrier may last for years. Also contact carrier is the term given to the carrier who has acquired the pathogen from the patient directly, whereas the *paradoxical* carriers are the ones who get infected from another carrier.

Animals: they are the reservoir hosts. When the infectious diseases are transmitted from animals to humans, it is termed as **zoonoses**, for example, plague caused by *Yersinia pestis* is present in rodents; rabies virus is spread to humans by a dog bite; protozoans and helminths (the common parasites) also have animals as their reservoir before infecting humans.

Insects: most of the members of Arthropoda are responsible for insect-mediated transfer of a disease. Insects such as mosquitoes, flies, fleas and lice are common vectors. Vectors can be mechanical (as in the case of house fly transmitting typhoid fever, cholera, dysentery, etc.) or it can be a biological vector. The pathogen multiplies inside the biological vector, completing a part of its life cycle before invading the host, e.g., *Plasmodium* is spread *via Anopheles* mosquito, and *Aedes aegypti* mosquito is the biological carrier for yellow fever.

Soil and water: soil contains many pathogens as they provide a medium for survival during the unfavorable conditions of growth. For example, *Clostridium tetani* spores, *Histoplasma capsulatum*, *Nocardia asteroides*, roundworms and hookworms are present in soil. *Vibrio cholerae* (causing cholera) and *Hepatitis* virus are water-transmitted pathogens.

Food: food poisoning is mainly due to staphylococcal infection.

Air: pathogens suspended in air in either droplets or dust are transmitted through airborne transmission. These pathogens are transmitted by coughing or sneezing.
    The course of the disease can be divided into several phases:
a) Incubation period: the time lag between the entry of the host to the appearance of first symptom
b) Prodromal stage: complete onset of signs and symptoms
c) Illness period: severity of disease eliciting immune response of the host
d) Convalescence: recovery phase

The step-wise infection process can be as follows:
1) Reservoir: to live before and after the infection
2) Transportation to the host
3) Attachment to the host, colonization and invasion
4) Multiplication inside the host cell
5) Escaping the host's defense mechanism
6) Causing harm to the host
7) Returning to the reservoir

These factors are discussed as under:

Reservoir: Most common reservoirs for pathogens are humans, animals and the nature itself.

Transportation: From the reservoir, the pathogen needs to be transported to the host, either through direct contact (coughing, sneezing and direct contact) or indirectly through inanimate objects (also known as fomites) such as soil, water and food. Inanimate objects for pathogen transmission are called as vehicles; for example, surgical instruments, bedding and utensils. The living transmitters of pathogen are called as vectors. They are also a mode of transmission for pathogens.

Adherence (adhesion, attachment) and colonization: It is the process by which bacteria gets attached to the surface of the host cells. It is a major step for the initiation of infection. Colonization basically refers to the pathogen establishment inside the host. Adhesins are the specialized adhesion factors required by bacteria for attaching to the host. Some of the adhesins are fimbriae (filamentous structure in bacteria) and capsule (exopolysaccharide layer outer to cell membrane), inhibiting phagocytosis by host defense, teichoic acid and lipoteichoic acid (components of Gram-positive bacterial cell wall).

Invasion: It is the process by which pathogens enter the host tissue and spread in the body. Entry into the cells is possible after penetrating the host's mucous membrane and epithelial cells. This is done by producing lytic substances which degrade the cell membranes. The pathogen can also enter through any crack or lesion in the body. Once inside the host, pathogen produces **virulence factors** (Table 2.4), and it also enters the bloodstream *via* the lymphatic capillaries surrounding the epithelium. Now they have the access to any of the organs of the host through the circulatory system.

Multiplication inside the host cell: for successful growth and reproduction inside the host, the pathogen must be present in an optimum environment. These intracellular pathogens have acquired mechanisms to fetch maximum nutrients and survive in the host. Few pathogens enter the bloodstream and this condition is known as **bacteremia**. This situation of pathogens in blood and their toxins in the bloodstream is called as **septicemia**. **Viremia** is the presence of viruses in the bloodstream.

Table 2.4: Virulence factors.

Factor	Pathogen	Mode of action
Coagulase	*Staphylococcus aureus*	Blood clot preventing the pathogen from phagocytosis
Collagenase	*Clostridium* spp.	Dissolves collagen of host tissues so that it can spread easily
Hemolysins	*Staphylococcus, Streptococcus, E. coli*	Weakening the immune system by anemia
Hydrogen peroxide and ammonia	*Mycoplasma* spp. and *Ureaplasma* spp.	Toxic to the epithelium of respiratory and urogenital systems
Porins	*Salmonella typhimurium*	Inhibit leukocyte phagocytosis

Exiting the host: this leaving of the host is also an important step as pathogen must now infect a new individual or host for spreading the disease. The exit can be active or passive. Active exit occurs when pathogen normally leaves the host through any exit point. Passive exit occurs when the pathogen exit is mediated *via* feces, urine, droplets, saliva, etc.

Toxicity to the host: the pathogen now multiplies inside the host and exchanges its genetic material mainly by transfer of plasmids. Many virulence factors coding genes are present on plasmids. **Virulence** is the degree to which the pathogen can cause damage to the host. Transposons also transfer virulence factors present on bacterial chromosome and plasmid DNA called "pathogenicity islands" between members of the same species or different species by horizontal gene transfer [20]. Many bacteria such as *Yersinia, Pseudomonas aeruginosa, Shigella, Salmonella* and *E. coli* have a minimum of one pathogenicity island.

A toxin is the substance (any metabolic product) that alters the normal metabolism of host cell [18]. The toxin produced by *Clostridium botulinum*, i.e., the botulinum toxin, is very harmful to the level (up to nanogram per kilogram concentration) that it itself eliminates the pathogen leaving only the symptoms of the disease [55]. **Toxigenicity** is the pathogen's ability to produce toxins, whereas **intoxications** are diseases that result from a specific toxin produced by the pathogen. Example of intoxications are such as botulinum toxin and aflatoxin (produced by mushrooms). Toxins can be divided into two groups: exotoxins and endotoxins. **Exotoxins** are heat labile (i.e., inactivated at 60–80 °C), soluble proteins that are released by the pathogen in their vicinity. They are produced mainly by Gram-positive bacteria and rarely by Gram-negative ones. These toxins after production move from their site of infection to other body parts. The genes for exotoxin production are normally present on plasmids. They are protein in nature and are easily recognized by the host immune system which produces immune antitoxins (antibodies that recognize and bind to toxins).

They can be converted to nontoxic **toxoids** after treating with chemicals such as form-aldehyde and iodine. The tetanus vaccine is a solution of tetanus toxoid. They can be further characterized as neurotoxin (affecting nervous tissues), enterotoxins (affecting intestines) and cytotoxins (affecting general tissues). Leukocidins and hemolysins are those which lyse WBCs and RBCs, respectively. They are also called as **superantigens** when they stimulate overexpression of T cells by binding to major histocompatibility complex (MHC) and T-cell receptors (TCR) for releasing massive amount of signaling molecules (cytokines) from other immune cells [34]. These cytokines cause multiple host organs to fail. Diseases resulting from exotoxins are botulism, diptheria and teta-nus. **Endotoxins** are the lipopolysaccharide (LPS) toxins produced by Gram-negative bacteria. Lipid A is the toxic component of LPS present in cell wall of Gram-negative bacteria. These are heat stable and mildly toxic in nature. They usually produce symp-toms such as fever, blood coagulation, diarrhea, inflammation and fibrinolysis. No toxoids can be formed for endotoxins and therefore no vaccines are available [53]. Diseases caused by endotoxins include meningococcemia and sepsis by Gram-nega-tive rods. **Mycotoxins** are toxins produced by fungi as secondary metabolites, for ex-ample, aflatoxin by *Aspergillus flavus* and satratoxins by *Stachybotrys*. Mycotoxins produce hallucinations, cancer and immunosuppression.

Evading the host defense mechanisms: pathogens prefer to bypass host defenses rather than killing the host. Capsules in the bacterial structure prevent complement activation, and the O side chain is enlarged in some gram-negative rods to prevent complement activation. The phagocytic action of the host is also altered as in the case of *Streptococcus pneumoniae*, *Neisseria meningitidis* and *Haemophilus influen-zae* by producing mucoid capsules or by producing specialized surface proteins (M protein) on *S. pyogenes*. These M proteins restrict the attachment of pathogen to the phagocytic cells (neutrophils, macrophages and monocytes). Leukocidins produced by *Staphylococcus* destroys WBCs. If they could not resist phagocytosis, then patho-gens develop methods to survive inside the phagocytic cells [15]. *Listeria monocyto-genes*, *Shigella* and *Rickettsia* move out from the phagocytic cells before they are merged with the lysosome for lysis. *Mycobacterium tuberculosis* becomes resistant to the lysosomal enzymes mainly because of the excess mycolic acid in its cell walls.

Virulence is measured in terms of **lethal dose 50**, which is the dose or number of pathogens that kills 50% of pathogens during a certain time.

2.4 Immune responses to bacterial infections

2.4.1 Immune mechanisms

If a pathogen attacks or invades anatomical barrier we say that infection has oc-curred. A healthy individual defends itself against pathogen by activating its immune mechanisms. The immune response can be of two types: **innate immunity** and **adaptive/acquired immunity**. The innate defense mechanism is the first line of defense while the adaptive immune system acts as the second line of defense. Both these defense mechanisms have cellular and humoral components by which they carry out their immune functions. Also, both these systems are interdependent upon each other, i.e., cells or components of the innate system influence the adaptive system and vice versa. The adaptive immune system requires some time to react to an invading pathogen, whereas innate system has mechanical and chemical barriers to react immediately against the invading pathogen. The adaptive system is a specific one and reacts only against those pathogens which elicited the response. This system also exhibits memory against the encountered pathogens. It recognizes the previously encountered pathogen and reacts more readily against the invading pathogen. The innate system does not have such kind of response. It is a type of nonspecific host defense mechanism against the invading bacteria. The most important components of innate immunity are anatomical barriers, complement system, and inflammatory and phagocytic responses. The enzyme lysozyme (antibacterial in nature) is present in mucus and all bodily secretions, and thus prevents bacteria from invading the internal tissues. If the pathogen enters the cell, then macrophages and neutrophils act upon it [46]. Complement is also activated to challenge bacterial infection.

Cellular defense: A variety of cells are involved in both kinds of immune systems. These cells include neutrophils [39], macrophages (involved in phagocytosis), basophils and mast cells (inflammatory response), B cells (antibody-mediated immunity) and T cells (cell-mediated immunity). All these cells are originated from the bone marrow. The hemocytoblast (stem cell in bone marrow) gives rise to myeloid progenitors which further generate neutrophils, eosinophils, basophils, monocytes and dendritic cells. Hemocytoblast also gives rise to lymphoid progenitor cells which further produce B cells and T cells. Monocytes give rise to macrophages and dendritic cells. B cells are released into the bloodstream and lymphatic system after their production in bone marrow. The plasma cells that secrete antibodies are developed through B cells. T-cell precursors undergo differentiation in thymus (hence, its name) into two distinct T-cell types: $CD4^+$ T helper (Th) cells and the $CD8^+$ cytotoxic T (Tc) cells. Macrophages and dendritic cells have the task of presenting antigens to the T cells for immune responses to take place [26].

The barriers of innate and adaptive immunities work against the bacterial infection.

2.4.1.1 Immune defenses against extracellular bacteria

Infections caused by **extracellular bacteria** are the common ones. In this case, the immune mechanisms are mainly due to the host's natural barriers, innate immune response and antibody protection. Skin and mucus layers are the first line of defense which prevent penetration and attachment of bacteria to the host. The mucociliar movement in the respiratory tract prevents the entry of pathogens. Acidic pH of the stomach kills bacteria entering the digestive tract, and salivary secretions contain antimicrobial chemicals such as lysozyme. The natural barriers against infection can be summarized as follows [38]:

a. Innate immunity: extracellular molecules (C-reactive proteins or CRPs, complement), NK cells, neutrophils, macrophages, chemokines and cytokines

b. Acquired immunity: antibodies and cytokines produced by T cells

Innate immune mechanism responds by actions of phagocytic cells, complement activation (alternative pathway), chemokine and cytokine production. When bacteria *Neisseria meningitidis* attack the host, they are acted upon by **complement proteins**. **Complement system** is a major innate immune response mechanism formed up of heat-labile plasma proteins. Mostly involved proteins are B, C1–C9 and D. These plasma proteins are normally present in inactive state circulating in blood and tissue fluids. They complement the immune system by enhancing the antibacterial activity of antibodies and as a result wiping out pathogens. These proteins react among themselves and initiate a cascade of enzymatic reactions generating proteolytic fragments and assisting in bacterial killing *via* three pathways: the classical complement pathway, the alternative complement pathway or the lectin pathway. These pathways lead to C3 activation, which further cleaves into a large fragment C3b (covalently attaches to pathogen's surface and acts as opsonin) and smaller C3a (anaphylotoxin) promoting inflammation by activating mast cells, releasing histamine. Activated C3 further damages the plasma membranes of bacteria by following a series of reactions.

The **classical pathway** requires binding of antibodies to the target pathogen. It involves complement components C1, C2 and C4. C1 is composed of C1q, C1r and C1s. The antigen–antibody complex binds to C1q by the Fc portion of the antibody. This complex is further joined by C1r and C1s which cleaves C4 and C2 forming C4b (larger fragment), C4a (smaller fragment), C2a (larger fragment) and C2b (smaller fragment). C3 convertase (C4b2a) is formed, which cleaves C3 into a large fragment C3b and smaller C3a. C3b forms a trimolecular complex with C3 convertase forming C5 convertase (C3b4b2a). C5 is broken into C5a and C5b by C5 convertase. Larger fragment C5b along with C6, C7, C8 and C9 binds to the surface of bacterium forming membrane attack complex, C5b6789 (MAC), or form opsonins that recognize bacterium for destruction. MAC can insert into the cell membrane of Gram-negative, but not Gram-positive, bacteria. In Gram-positive bacteria, it produces pores that allow the entry of membrane damaging molecules, such as lysozyme, and makes the bacterium susceptible to osmotic lysis. Complements C3a and C5a have other functions which are discussed later.

The **alternative pathway** is triggered by bacterial endotoxin and other LPSs found on the surface of attacking pathogen. Antibody is not required in this pathway for lysis of bacteria. This pathway involves factors B, D, H and I. C3 undergoes spontaneous hydrolysis and with the help of factors B and D forms C3b. This further forms C3 convertase (C3bBb) of alternative pathway, activating a cascade leading to formation of MAC causing lysis of bacterium.

During the **lectin pathway**, mannose-binding lectins (MBL) synthesized in the liver bind to proteins containing glucose and mannose residues that are found in cell wall of *Neisseria*, *Candida* and *Salmonella* spp. Once bound, MBL forms a complex with an enzyme called MBL-activated serine protease. In this form, this enzyme activates C3 convertase and C4b2a (by cleaving C2 and C4 complement components) that participates in forming MAC [22].

Another mode of action is **phagocytosis**. Factor C3b activates neutrophils and macrophages which engulf foreign bacteria or antigen by the process known as **opsonization**. These opsonized bacteria are then recognized and destroyed by phagocytic cells. Opsonization allows killing of Gram-positive bacteria (e.g., *Staphylococcus* spp.) that are resistant to killing by MAC. After bacteria are ingested by phagocytosis, they are destroyed and broken into small fragments by enzymes. Phagocytes present the fragments on their surface *via* MHC class II molecules. Circulating Th cells recognize these bacterial fragments and begin to produce proteins called cytokines. Th1 and Th2 are two groups of Th cells differing in the type of cytokine they secrete. Th1 cells principally produce interferon-g (IFN-g), which promotes cell-mediated immune mechanisms. Th2 cells produce mostly interleukin (IL)-4, which promotes humoral immunity by activating B cells. B cells produce antibodies that stick to extracellular bacteria and prevent their growth and survival.

Complement C5a guides neutrophils and macrophages to the area where antigen or foreign bacteria are present by the process of **chemotaxis**. Complements C3a, C4a and C5a activate mast cells and basophil which release histamine and serotonin. Histamine and serotonin produce inflammatory reactions.

2.4.1.2 Immune defenses against intracellular bacteria

Some pathogens after entering the phagocytic cells can evade killing mechanisms to survive inside cells [49]. The **intracellular bacteria** (e.g., *Salmonella* spp.) target macrophages. These bacteria because of living inside the cell cannot be acted upon by complement or antibody but, instead, are abolished using a **cell-mediated response**. Infected macrophages present the bacterial peptide using MHC II on their cell surface, called as **antigen presentation**. A Th cell with its TCR releases IFN-g. This cytokine stimulates killing mechanisms (such as production of lysozyme) inside the infected macrophage to digest and destroy the invading bacterium. IFN-g also increases antigen presentation by cells, making the bacterium more visible to the immune system and more prone to attack. The intracellular pathogen has the main challenge of surviving within the macrophage, e.g., *Mycobacterium tuberculosis*, *M. leprae* and *Listeria monocytogenes*. Entering the host cell is also a barrier for a pathogen. Bacterial penetration elicits a huge inflammatory immune response in the host. When inside the macrophages, *Mycobacterium* stimulates the MHC II-presented antigen to CD4$^+$ Th cells or MHC I-presented antigens to CD8$^+$ Tc cells. Activated CD4$^+$ T cells secrete IFN-gamma, which further activate macrophages leading to NO production and pathogen destruction [61]. The CD8$^+$ cells are cytotoxic in nature and destroy the pathogen by destroying the macrophages. *M. tuberculosis* pathogen cannot be completely eradicated from the host.

The CRP is a pentameric protein produced by hepatic cells of the liver in response to bacterial infections. CRP often binds to phospholipid in the membrane of some bacteria such as *Pneumococcus* and functions like opsonin, facilitating phagocytosis by neutrophils. Inflammation causes formation of IL-6, which goes to the liver, and then the liver forms CRP. CRPs are also known as acute phase reactants. CRP binds to the phosphocholine receptors present upon bacteria-activating macrophages. Macrophages kill the bacteria, and the smaller fragments of killed bacteria further activate the C1q of a classical complement pathway. Complement activates the MAC (C5–C9) and facilitates opsonization through C3b component of complement. Serious infections by *Neisseria meningitides* and *Neisseria gonorrhea* occur if the complement system is inefficient. The mast cells and basophiles are activated by complement as in the case of C5a, C3a and C4a and release mediators which after combining with these complement proteins entice leukocytes at the site for the necessary action. CRP also stimulates the synthesis of tumor necrosis factor (TNF)-alpha, thus inducing the synthesis of NO and consequently the destruction of various microorganisms. **Eosinophils** are bacteriocidal in action, and their activity is studied against *E. coli*. A rapid and efficient killing of pathogen was observed under aerobic conditions, whereas under anaerobic conditions the killing was not efficient. Neutrophils and macrophages produce NO and hydrogen peroxide to which bacteria are susceptible. Neutrophils have a short life span, whereas macrophages have a comparatively longer life span. Neutrophils during inflammation produce pus-filled secretions, whereas macrophages form granuloma [56]. Neutrophils mainly protect against extracellular bacteria, whereas macrophages act against the cells harboring intracellular bacteria. **Cytokines** are the signaling molecules produced by immune system cells for specific functions [10], for example, IL is a type of cytokine produced by white cells as signaling molecules. **Chemokines**, the small cytokines, are signaling proteins produced by the cells. They can attract other defensive cells at the site of infection. TNF-alpha, IL-1 and IL-6 are the prominent cytokines taking part in action against bacteria. These are produced in the initial phase of infection. They act upon hypothalamus causing fever, inhibiting bacterial multiplication. They also increase expression of selectins (family of cell adhesion molecules, CAM), thereby facilitating cells to the site of infection. Cytokines also stimulate neutrophils and macrophages to produce NO that destroys bacteria. There are other cytokines also which interfere with adaptive immune response. **IL-2** is produced by macrophages, and dendritic cells in response to pathogenic stimulation [24]. It has a function of differentiating naïve T cells to Th1 cells, whereas **IL-4** produced by basophiles, mastocytes and macrophages, stimulates a differentiation of naïve T cells into Th2 cells, which helps B cells to produce antibodies, mainly IgE. **Antibodies** acting under the influence of adaptive immune system responds to bacterial infection mainly by opsonization, complement activation and neutralization of bacteria and its products. Antibodies also facilitate phagocytosis by neutrophils and macrophages. These phagocytic cells have receptors for Fc portion of immunoglobulins, whereas the Fab portion of antibodies bind to the antigens, thus bringing the phagocytosis of pathogen. Antibodies also coassist in destroying bacteria by the complement and activate this system by a classic pathway. IgA binds with the bacteria and restricts them from binding to the mucus membrane, intestinal tract and respiratory act. Antibodies can also bind to bacterial toxins (e.g., tetanus and diptheria toxins) and can neutralize them. Gram-negative bacterial infections result in septicemia and septic shock. Septic shock is due to the LPS layer in the outer membrane of cell wall. LPS leads to an intensified production of proinflammatory cytokines in the neutrophils, macrophages, endothelial cells and muscles (TNF-alpha, IL-1, IL-6, IL-8) and NO. Due to this, the muscle tone and heart beat are reduced, leading to hypotension and further cellular death.

2.5 Chemotherapy: control of bacterial diseases and diagnosis

2.5.1 Chemotherapy

Chemotherapy is the use of chemical drugs to treat any disease. Chemotherapeutic agents are used to treat the diseases, the dosage of which should be lower enough to combat pathogen and still being harmless to the host. Antibiotics are the commonest chemotherapeutic agents used, which can be microbial products or chemicals that either kill the pathogen or check the growth [19]. Antibiotics can be chemically synthesized, for example, sulfonamides. During the 1920s, German physician Paul Ehrlich introduced the first antibacterial drugs to fight diseases who also received the Nobel Prize (with Elie Metchnikoff) a century ago for his work on immunity. In 1904, Ehrlich found that the dye trypan red was active against the trypanosome that causes African sleeping sickness and could be used therapeutically. Ehrlich successively established chemotherapy of sleeping sickness and syphilis and gave the concept of selective toxicity [7]. In 1927, Gerhard Domagk along with the chemical industry owner I. G. Farbenindustrie discovered prontosil red, a dye used to stain leather which was nontoxic for animals and protected mice against streptococcal and staphylococcal infections. In 1935, it was established that pronto-sil red actually converted into an active factor called sulfonamide by French scientists Jacques and Therese Trefouel. Domagk had discovered sulfonamides or sulfa drugs and received the Nobel Prize in 1939.

Penicillin was discovered by the Scottish physician Alexander Fleming. He was working on wound infections since the First World War and was trying to find something that would be harmful for pathogens. He left a Petri plate inoculated with staphylococci exposed on his working bench, which accidentally got contaminated with a species of *Penicillium notatum*. Fleming then noticed after few days that *Penicillium* colony was growing at one end of the plate and *Staphylococcus* on the other; he then stated that the fungus *Penicillium* was producing a diffusible substance which was lethal to staphylococci. He called this diffusible substance as penicillin. But Fleming somehow could not be convinced that penicillin remain active in the body long enough after injection to destroy the pathogens and he decided to drop the research [21]. In 1939, Howard Florey (pathology professor at Oxford University) and his coworker Ernst Chain who were working on testing of the bacteriocidal activity of lysozyme and sulfonamides purified penicillin by asking for the *Penicillium* culture from Fleming. Along with the help of biochemist Norman Heatley, they purified crude penicillin and injected it into mice infected with *Streptococcus* and *Staphylococcus*. They found that the mice survived the infections. Fleming, Florey and Chain received the Nobel Prize in 1945 for the discovery and production of penicillin. In 1944, Selman Waksman discovered streptomycin produced by *Streptomyces griseus* (an actinomycete). He received Nobel Prize for the same in 1952.

2.5.1.1 Antimicrobial drugs

The antimicrobial drugs must have selective toxicity, i.e., it should be harmful only to the pathogen and not to the host. The amount of drug required for clinical treatment of any infection is called its therapeutic dose, and the drug amount administered at which it becomes lethal for the host is called as toxic dose. Any drug that targets the microbial function not found in eukaryotes has a higher therapeutic index (ratio of toxic dose to therapeutic dose) and greater selective toxicity. The higher the therapeutic index the better the drug. Penicillin blocks the cell wall synthesis mechanism in bacteria, but the host cells lack cell wall so has a negligible effect on the host and has a high therapeutic index [5]. Drugs can be narrow spectrum (effective against a limited variety of pathogens) or broad spectrum (active against a wide variety of pathogens). These antimicrobial chemicals can be obtained naturally or can be manufactured synthetically. The example of naturally produced antibiotics are amphotericin B, chloramphenicol, erythromycin, kanamycin, neomycin, nystatin, rifampin, streptomycin, tetracyclines and vancomycin produced by *Streptomyces* spp. Bacitracin and polymyxins are produced by *Bacillus* spp. and fungus *Penicillium* and *Cephalosporium* produced penicillin and cephalo-sporin, respectively. Sulfonamides, chloramphenicol and ciprofloxacin are the examples of synthetic drugs. Some antibiotics are chemically modified to make them less susceptible to inactivation, and such antibiotics are called as semisynthetic, for example, ampicillin, carbenicillin and methicillin. The chemotherapeutic agents can either be bacteriocidal or bacteriostatic; cidal refers to killing of bacteria and static refers to checking the growth of the bacteria. Bacteriocidal drugs kill bacteria which are then removed by hosts' immune defense system. Whereas the static ones check the growth of pathogens, but their dosage must be taken care of. If the drug is insufficient, the pathogen may again grow and can cause infection. Examples of bacteriostatic drugs are chloramphenicol, erythromycin, novobiocin, sulfonamides and tetracyclines. Some bacteriocidal drugs are ampicillin, carbenicillin, cephaloridine, cloxacillin, colistin, gentamicin, kanamycin, lincomycin, methicillin, penicillin, polymyxin and streptomycin [16].

Minimum inhibitory concentration is the minimum amount of antibiotic required to inhibit the growth of the pathogen, whereas **minimum lethal concentration** is the minimum amount of drug required for killing of the pathogen.

These bacteriocidal chemotherapeutic drugs act mainly by any of the three mechanisms: *inhibiting cell wall synthesis*, disturbing the cells' osmolality and causing the cell to absorb more water and then lyse. They also affect the *cell membrane synthesis* (resulting in loss of metabolites) and *protein synthesis*. The bacteriostatic agents interfere with protein synthesis at an early stage and hence prevent the growth and proliferation of bacteria rather than killing them. The major sites of action of antibiotics are given in Table 2.5.

Table 2.5: Mechanism of action of antibiotics.

Drug	Mode of action
Inhibition of cell wall synthesis	
Penicillin	Inhibit synthesis of the bacterial cell wall peptidoglycan
Ampicillin	Activate cell wall lytic enzymes
Carbenicillin	
Methicillin	
Cephalosporins	
Vancomycin	
Bacitracin	
Inhibition of protein synthesis	
Streptomycin	Binds to the ribosomal subunits to inhibit translation
Gentamicin	
Chloramphenicol	
Tetracyclines	
Erythromycin	

Table 2.5 (continued)

Drug	Mode of action
Inhibition of nucleic acid synthesis	
Ciprofloxacin	Blocks replication, transcription
Rifampin	Blocks transcription
Damaging the cell membrane	
Polymyxin B	It binds to the plasma membrane and disrupts its structure
Antagonism	
Sulfonamides	Inhibit folic acid synthesis by competition with p-aminobenzoic acid
Trimethoprin	Blocks tetrahydrofolate synthesis through inhibition of the enzyme dihydrofolate reductase

Streptomycin, gentamicin, spectinomycin, clindamycin, chloramphenicol, tetracyclines, erythromycin and many other antibiotics inhibit translation by binding with the ribosomal subunits (50S and 30S). The therapeutic indices for such kind of drugs are also high because they are able to differentiate between prokaryotic and eukaryotic ribosome and targeting only prokaryotic ribosomes. Chloramphenicol interferes with the mRNA attachment step to the ribosome; erythromycin interferes with the attachment of amino-acyl tRNA complex to the ribosome. Streptomycin, kanamycin and neomycin cause distortion of the 30S subunit of ribosome, resulting in faulty translation.

As stated earlier, the drugs interfering with cell wall synthesis are considered more effective as this cell wall is not found in eukaryotic host, causing a minimal harm to the host. The other drugs having mode of actions such as cell membrane disruption or nucleic acid synthesis alteration are not that effective because of the similar mechanisms in both prokaryotes and eukaryotes.

Penicillin prevents the cross-linking occurring in the cell wall by transpeptidation. The cell wall of bacteria contains mainly peptidoglycan, teichoic acids, D-amino acids and diaminopimelic acid which are absent in eukaryotes and hence the hosts' cells are resistant to penicillin. The bacterial mucopeptidases lyse the inner part of cell wall in the dividing cell so that the cytoplasm expands and meanwhile the exterior part of cell wall is enlarged. Penicillin is not effective in low concentrations as the cell continues to produce mucopeptidases. Thus, it takes time for penicillin to act upon the bacteria. Chloramphenicol hampers the mucopeptidase synthesis which is important to completely block cell wall synthesis. In higher concentration, penicillin is quite effective. The absence of cell walls in mycoplasma is responsible for their penicillin resistance. Cycloserine, an analog of D-alanine (an amino acid in the side chain of cell wall), also interferes with cell wall synthesis.

Amphotericin B and nystatin bind to sterol of cell membrane and interfere with osmotic activity of the cell. Quinolones and polymyxins affect nucleic acid synthesis or membrane structure.

Some drugs are antimetabolites as they block the metabolic pathways by inhibiting the enzymes. Sulfonamides (sulfanilamide, sulfamethoxazole, sulfacetamide) also have a high therapeutic index as they inhibit folic acid synthesis and humans cannot synthesize folic acid, and this acid is essential for their diet. The pathogens synthesize their own folic acid. In many cases of infection, combination of drugs is used for treating the disease. Penicillin is given in combination with streptomycin for the treatment of endocarditis (caused by enterococcal infection) and bronchial infection (caused by *Haemophilus influenzae*).

2.5.1.2 Drug resistance

Drug resistance is one of the serious threats for microbial disease treatment. This resistance can develop by any of the three methods: natural drug resistance (the organism is never exposed to the drug), acquired drug resistance (occurred during treatment) and transferred drug resistance (genetic transfer). They can also prevent the entry of drug. Most of the Gram-negative pathogens are resistant to penicillin as the drug cannot cross the outer membrane of the cell. Mycobacteria have high content of mycolic acid outside the peptidoglycan in the cell walls and hence are resistant to most of the drugs. Some pathogens banish drugs out of the cell by their efflux pumps (plasma membrane translocases), and this phenomenon is called as multiple drug resistance [27, 37, 43]. Such kind of drug resistance are present in *E. coli*, *Pseudomonas aeruginosa*, *Mycobacterium smegmatis* and *Staphylococcus aureus*. Another category of pathogens are those that chemically modify drugs to inactivate them; for example, penicillin is inactivated by enzyme penicillinase, which hydrolyzes the lactam ring present in the penicillin structure. Chloramphenicol is acetylated, and the aminoglycosides are acetylated or phosphorylated by the pathogens for inactivation. Resistance can also be achieved if the target of the drug is modified; for example, erythromycin and chloramphenicol act on ribosomes, if the 23S rRNA is modified they cannot bind to the ribosomes and therefore cannot be harmful for the pathogen. Enterococci become resistant to vancomycin by changing the terminal D-alanine–D-alanine in their peptidoglycan to a D-alanine–D-lactate. The tuberculosis (TB) bacteria *Mycobacterium* develop resistance against the drug rifampin by altering the beta subunit of RNA polymerase by mutation; thus, rifampin acts by blocking the transcription initiation by RNA polymerase. The sulfonamide-resistant bacteria use preformed folic acid rather than synthesizing the chemical or they just increase the folic acid synthesis rate for developing drug resistance. The drug resistance genes are found on bacterial chromosome and plasmids. Spontaneous mutations also cause the drug resistance in bacteria. These mutations change the drug receptor; therefore, the antibiotic cannot bind. The plasmids containing the drug resistance genes are called **R plasmid** (resistance plasmid). These genes code for enzymes that destroy drugs. Plasmid-associated genes impart resistance to the aminoglycosides, choramphenicol, penicillins and cephalosporins, erythromycin, tetracyclines, sulfonamides and others. These plasmids are transferred to other bacteria by processes such as conjugation, transformation and transduction. Also, it must be noted that more than one drug resistance genes can be present on the same plasmid. **Superinfection** is caused when the pathogen develops resistance against multiple drugs, for example, disease pseudomembranous enterocolitis caused by *Clostridium difficle*. Among patients treated with antibiotics such as clindamycin, ampicillin or cephalosporin, it is seen that many intestinal bacteria are killed, but *C. difficile* survive the drug treatment causing pseudomembranous enterocolitis. *C. difficile* is a minor member of the intestine but due to the absence of its competitors it grows in number and produces a toxin. This toxin stimulates the secretion of pseudomembrane by intestinal cells. The treatment for this is possible by vancomycin; otherwise, the pseudomembrane needs to be surgically removed. Fungi *Candida albicans* also produce superinfections when bacterial competition is eliminated by antibiotics.

Therefore, the antibiotics should be administered in ways that minimize the development of resistance. Also new antibiotics could be researched and discovered in order to fight the problem of drug resistance.

2.6 Systemic bacteriology and bacterial diseases

2.6.1 Systemic bacteriology

The term systemic refers to the spread of infection throughout the systems of body, whereas septicemia is the presence of bacteria in blood. This type of infection affects a larger part of body. Systemic bacterial infection is sometimes called as sepsis. All the clinically important bacteria come under the systemic bacteriology. The pathogen is distributed throughout the body, and the toxins produced by them are the major reason behind the symptoms which include cold, flu, sore throat, chills, fever, nausea, vomiting and weakness. A number of bacteria live in the human gut, providing a symbiotic relationship with the host. But if they cross the intestinal epithelium, they enter the blood causing infection; for example, *E. coli* and *Citrobacter rodenticum* are present in the intestine but they can also produce a toxin lymphostatin in blood [9]. This kind of intestinal injury can be treated by specific antibodies against lymphostatin for treating and preventing extraintestinal problems of Gram-negative infection.

Some of the bacterial diseases along with their mode of transmission are given further.

2.6.1.1 Airborne disease

2.6.1.1.1 Diphtheria
It is a contagious disease caused by the Gram-positive *Corynebacterium diphtheriae*. It is transmitted by nasopharyngeal secretions and is resistant to dry conditions. Symptoms include the presence of mucus and pus in nasal discharge. The diagnosis is done by pseudomembrane examination formed in the throat and by examining the bacterial culture. Penicillin and erythromycin are used for treatment. Immunization with the **DPT** (**d**iphtheria–**p**ertussis–**t**etanus) **vaccine** is a preventive measure. *C. diphtheriae* can also infect the wounded skin or at skin lesion, causing a slow-healing ulceration termed **cutaneous diphtheria.**

2.6.1.1.2 Legionnaires' disease and Pontiac fever
Legionella pneumophila, a Gram-negative rod, is present in the environment and infects the human respiratory system. Males with over 50 years of age who are immunocompromised due to heavy smoking, alcohol consumption and any chronic illness develop the disease. The symptoms for the disease are high fever, cough without secretions, headache, neurological disorders and bronchopneumonia. Diagnosis is done by isolation of bacterium and checking the antibody titer or by a kit using urine for detection of antigens. Erythromycin and rifampin are advised to the patients. *L. pneumophila* also causes an illness called **Pontiac fever**. Symptoms of Pontiac fever include onset of fever, headache and muscle pains.

2.6.1.1.3 Meningitis
The brain or spinal cord meninges are infected. The causative pathogen initially enters through nasopharynx after which they enter the blood circulation and then CSF (cerebrospinal fluid) by crossing the mucosal barriers. Symptoms include respiratory illness, sore throat, vomiting, headache, fatigue, confusion, and neck and back stiffness. Diagnosis of bacterial meningitis can be done by Gram stain and culture of the causative bacteria from CSF [51]. Once it is diagnosed, penicillin, chloramphenicol, cefotaxime, ceftriaxone and ofloxacin are advised.

2.6.1.1.4 Pertussis
Pertussis is called "whooping cough." This is caused by the Gram-negative bacterium *Bordetella pertussis*. It is a highly contagious disease mainly affecting children. Transmission occurs by the release of droplets from the infected person [17]. After the pathogen's entry into the respiratory tract, the bacteria produce adhesins (hemagglutinin) and attach themselves to the epithelial cells. Laboratory diagnosis is done by culturing the bacterium, fluorescent antibody staining of smears from

nasopharyngeal swabs and serological tests. Vaccination is done by DPT vaccine, and children have to be vaccinated when they are 2–3 months old.

2.6.1.1.5 Streptococcal diseases

Streptococcus is a Gram-positive bacterium, widely distributed among humans (mainly asymptomatic carrier) [1]. Transmission of the pathogen occurs by respiratory droplet, direct or indirect contact. *Streptococcus pyogenes* is the common pathogen infecting humans. Diagnosis can be done by both clinical and laboratory tests. Antibiotic penicillin or erythromycin is given for its treatment. The control of streptococcal diseases is difficult as vaccines are not available for them except for streptococcal pneumonia because of the large number of serotypes (serologically distinguishable strains of *Streptococcus*). Best way to stop infection is to control the transmission. Infected individuals should be kept in isolation and treated. Some streptococcal diseases are as follows:

Cellulitis and erysipelas: it is the infection of deeper layer of the skin tissue. The inflammation is characterized by erythema (redness on skin) and edema (accumulation of fluids). **Impetigo**, caused by *S. pyogenes* and *S. aureus*, is a superficial infection of cutaneous tissues of face. Penicillin or erythromycin is administered for infection. **Erysipelas** is infection of dermal layer of the skin occurring mainly in infants. Patients develop shiny red patches or lesions that enlarge and thicken with a sharply defined edge. Penicillin and erythromycin are given for treatment.

Poststreptococcal diseases: they occur 1–4 weeks after acute streptococcal infections. The diseases are glomerulonephritis and rheumatic fever. **Glomerulonephritis** or **Bright's disease** is an inflammatory disease of the renal glomeruli, membranous structures within the kidney where blood is filtered. The disease is due to type III hypersensitivity reaction, involving deposition of antigen–antibody complex. The complexes cause destruction of the glomerular membrane, leading to blood and protein leakage in urine. Symptoms include edema, fever, hypertension, and hematuria (blood in the urine). Diagnosis is done by clinical history, physical findings and study of prior streptococcal infection as there are no specific laboratory tests available. Antibiotic penicillin G or erythromycin can be recommended to control streptococci. There is no specific therapy once kidney is damaged.

Rheumatic fever: is an autoimmune disease, which causes permanent damage to the heart, damaged heart valves, subcutaneous tissue and CNS inflammation. It usually occurs after a prior streptococcal sore throat infection. The only way to cure is to decrease the inflammation and fever and control cardiac failure. Antibiotic penicillin may be given to remove the remaining streptococcus bacteria in the body. Naproxen, an anti-inflammatory drug, is given to reduce pain, inflammation and fever. Prednisone, a corticosteroid, is also recommended in case the preliminary medicines do not work or in cases of heart inflammation.

Scarlet fever: *S. pyogenes* carrying a lysogenic bacteriophage causes the throat infection. This codes for rash inducing toxin which causes shedding of the skin. The rashes are due to the generalized reaction to the toxin. It is a communicable disease, and the symptoms include sore throat, chills, fever, headache and a strawberry-colored tongue. Treatment is done with penicillin.

Streptococcal sore throat: also called as strep throat is the most common bacterial infections of throat including tonsils. It is caused by group A *Streptococcus* bacteria. Disease is spread by sneezing and coughing which releases droplets of saliva or nasal secretions in the air. The pathogen causes **pharyngitis** or **tonsillitis** stimulating inflammatory response and the lysis of leukocytes and erythrocytes. Symptoms include sore, red throat, fever, chills, headache, swollen lymph nodes and difficulty in swallowing. Rapid strep test is performed for diagnosing strep throat. Penicillin and amoxicillin (azithromycin or erythromycin for penicillin allergic people) are given for treatment. Infection is more common in children than adults. In adults, the disease tends to be milder and less frequent because of better developed immunity.

Streptococcal pneumonia: it is caused by the Gram-positive bacterium *Streptococcus pneumoniae* present in the upper respiratory tract. The capsular polysaccharide is the main virulence factor which is mainly composed of hyaluronic acid. This protects the pathogen from ingestion and getting killed by phagocytes. Pneumolysin is also produced destroying the host cell. The alveoli get filled with blood cells, and the sputum also gets rust colored because of blood in the cough from lungs. Symptoms include chills, hard-labored breathing and chest pain. Diagnosis is by chest X-ray, biochemical tests and culture. Penicillin G, cefotaxime, ofloxacin and ceftriaxone are recommended for treatment.

2.6.1.1.6 Tuberculosis

Mycobacterium tuberculosis is the causative agent of TB. The pathogen primarily affects the lungs. The bacterium comes in contact with macrophages and is phagocytosed. Infected macrophages in lungs produce chemokines attracting monocytes, lymphocytes and neutrophils to the site of infection. With the development of cellular immunity, macrophages are killed resulting in the formation of small hard nodules called tubercles (due to the hypersensitivity response). The *Mycobacterium* remain dormant inside the macrophages, as they cannot divide due to acidic pH, low oxygen level and other toxic fatty acids. Resisting oxidative killing, inhibition of phagosome, lysosome fusion and inhibition of diffusion of lysosomal enzymes are the mechanisms by which *Mycobacterium* survives inside macrophages [11]. Symptoms include chest pain, blood in sputum, weakness and weight loss. During the course of disease, the tubercles calcify and now are termed as **Ghon complexes**. These complexes are observed in the X-ray. Hosts develop cell-mediated immunity, which involves sensitized T cells. This forms the base for the tuberculin skin test. In this test, a purified protein derivative (PPD) of *M. tuberculosis* is injected intracutaneously (**the Mantoux test**). In definite hosts, the sensitized T cells react with PPD and show a delayed hypersensitivity reaction within 2 days. The hypersensitivity is visible as hardening and reddening at the site of injection. A multidrug-resistant TB (**MDR-TB**) is the type of TB caused by bacteria those are resistant to isoniazid and rifampin (two most effective drugs in treating TB) [29]. People who are HIV infected are more likely to develop TB than normal beings. Infants and children are vaccinated with **b**acille **C**almette-**G**uérin vaccine to prevent complications such as meningitis.

2.6.2 Arthropod-borne disease

These diseases are rare; ehrlichiosis, Q fever and Lyme disease are some of the examples.

2.6.2.1 Ehrlichiosis

Ehrlichia chaffeensis is the causal agent. It is transmitted to humans by biting of ticks such as lone star tick (*Amblyomma americanum*) and black-legged tick (*Ixodes scapularis*). *E. chaffeensis* invades monocytes and macrophages and grows inside vacuoles within these cells in blood and also in tissues like bone marrow, liver, spleen, kidney, lungs and CSF. This causes human monocytic ehrlichiosis, a nonspecific febrile illness that resembles Rocky Mountain spotted fever. Serological tests are performed for diagnosis, and tetracycline is given for treatment.

2.6.2.2 Lyme disease

This is a tick-borne disease. The causal organisms are *Borrelia burgdorferi*, *B. garinii* and *B. afzelii*. Deer and mice are the natural hosts. Within 10 days of infectious tick bite a ring-shaped skin lesion with red border is observed [4]. Symptoms include fatigue, headache, fever and chills. Severe symptoms develop after weeks or months: neurological abnormalities, heart inflammation and arthritis. Laboratory diagnosis of Lyme disease is done by performing polymerase chain reaction for finding *B. burgdorferi* DNA in urine, and serological tests for IgM and IgG antibodies. Amoxicillin or tetracycline are given to cure this disease.

2.6.2.3 Plague

This is caused by Gram-negative bacterium *Yersinia pestis* transmitted from rodent to humans [48] by the bite of an infected vector flea, or by direct contact with infected tissues and inhalation of infected respiratory droplets. Symptoms include fever and lymph node enlargement (**buboes,** hence the name, **bubonic plague**). **Septicemic plague** occurs when bacteria circulate in the bloodstream, and symptoms of which are fever, abdominal pain, diarrhea, bleeding from nose or mouth. **Pneumonic plague** occurs when *Yersinia* is reached to the lungs whose symptoms are chest pain, nausea, vomiting and difficulty in breathing. Laboratory diagnosis is made by taking samples from patient's blood and buboes and performing direct microscopic examination, culture of the bacterium, serological tests, detection of pathogen DNA in infected fleas and phage testing. Antibiotics such as streptomycin, gentamicin, chloramphenicol or tetracycline can be administered in case of the disease [41].

2.6.3 Direct contact disease

They are mainly concerned with the skin and underlying tissues.

2.6.3.1 Anthrax

It is transmitted to humans by direct contact with the infected cattle, sheep and goats. The pathogen is Gram-positive *Bacillus anthracis*. Symptoms include skin papule that ulcerates (**eschar),** headache, fever and nausea. Diagnosis is by direct microscopic examination, bacterial culture and serology. Penicillin G and streptomycin are given for its treatment. Vaccination of animals, mainly cattle, is an important control measure.

2.6.3.2 Gas gangrene or clostridial myonecrosis

Clostridium perfringens, *C. novyi* and *C. septicum* are Gram-positive rods causing the disease. They produce gas gangrene or clostridial myonecrosis, an infection of skeletal muscles. Infections are commonly associated with deep wounds, muscle injuries or wounds that are contaminated with stool and dirt, severely damaged tissues resulting from automobile accidents, military combat or frostbite. The pathogen secretes alpha toxin which dissolves muscle tissue. As a result, the accumulation of hydrogen gas (result of carbohydrate fermentation) occurs in the skeletal muscle tissue. Symptoms include blisters with foul smelling discharge, severe pain around the wound, edema and muscle necrosis. It can be treated by surgically removing the dead tissues or by administration of polyvalent antitoxin, penicillin and tetracycline.

2.6.3.3 Gonorrhea

This is a sexually transmitted disease (STD) of the mucus membrane of the genito-urinary tract, eye, rectum and throat. It is caused by the Gram-negative, oxidase-positive, diplococcus, *Neisseria gonorrhoeae*. These pathogens get attached to the mucosal cells by means of pili and protein II (adhesins) and hence are not washed away by normal vaginal discharge or by flow of urine [2]. They are phagocytosed by neutrophils and now they become intracellular, and host tissues respond by infiltration of mast cells, polymorphonuclear neutrophils and plasma cells. Symptoms include painful or swollen testicles; urethral discharge of yellow and creamy pus; frequent, painful urination followed by a burning sensation in males and in females; and more vaginal discharge [44]. Laboratory diagnosis of gonorrhea is done by culturing the pathogen for determining the oxidase reaction, Gram's staining and studying colony and cell morphology. Penicillin G plus probenecid, ampicillin plus probenecid, ceftriaxone or ofloxacin plus doxycycline and spectinomycin are recommended for treatment. Probenecid increases the antibiotic blood levels.

2.6.3.4 Leprosy

Leprosy is also known as **Hansen's disease** caused by bacillus *Mycobacterium leprae*. Transmission occurs mainly by exposure for prolonged periods to infected individuals shedding large numbers of *M. leprae*. The bacterium invades mucosa of upper respiratory tract, peripheral nerve and skin cells and becomes an obligate intracellular parasite. Target for the bacterium is mainly the Schwann cells surrounding the peripheral nerve axons and mononuclear phagocytes. The bacteria triggers Schwann cells and reverts them to stem cell like state. These cells now travel to other organs of body and the pathogen moves along with them. Symptoms include slightly pigmented skin eruption several centimeters in diameter, skin becomes thick and stiff. The leprosy pathogen cannot be cultured *in vitro*, so the diagnosis is based on biopsy studies and acid-fast staining. Serodiagnostic methods such as radioimmunoassay using *M. leprae* antigen, the fluorescent leprosy antibody absorption (ABS) test, DNA amplification, serum antibody competition test, inhibition assay using monoclonal antibodies against specific epitopes of *M. leprae* and ELISA have been developed. Treatment is possible with combination of antibiotics, also called as multidrug therapy such as sulfone drug diacetyl dapsone and rifampin, with or without clofazimine. Multidrug therapy prevents development of resistance in bacteria. Alternative drugs are ethionamide or protionamide.

2.6.3.5 Staphylococcal diseases

Staphylococcus are Gram-positive cocci, occurring singly, in pairs and in tetrads. Cell wall comprises peptidoglycan and teichoic acid. Staphylococci are facultative anaerobes, salt tolerant, hemolytic and are usually catalase positive. Strain *S. aureus* is coagulase-positive (ability to clot blood) and is more virulent and dangerous than coagulase-negative *S. epidermidis*. This pathogen is ubiquitous and resides in the upper respiratory tract, skin, intestine and vagina. Staphylococci, with pneumococci and streptococci, are pyogenic bacteria (pus-formers). These bacteria cause pus-forming diseases such as boils, carbuncles, folliculitis, impetigo in humans. It spreads by respiratory passages, by animate or by inanimate objects. Staphylococci produce disease through their direct tissue invasion and through their production of many extracellular substances (catalase, coagulase, hyaluronidase, enterotoxins, DNase, hemolysins, lipases, proteases and nucleases). The pathogen spreads from the lymphatics and bloodstream to other parts of the body. **Toxic shock syndrome** is caused by toxin (**t**oxic **s**hock **s**yndrome **t**oxin-1 is a superantigen) released by *Staphylococcus* and is characterized by low blood pressure, fever, diarrhea, an extensive skin rash and shedding of the skin [58]. **Staphylococcal scalded skin syndrome** is caused by strains of *S. aureus* that carry a plasmid-borne gene for the **exfoliative toxin** or **exfoliatin**. Exfoliatin is also a superantigen, and the disease is characterized by peeling off of epidermal layer and a red area beneath [47]. The final diagnosis of staphylococcal disease can be made only by isolation and identification of the pathogen involved. This can be done by culturing the pathogen and performing catalase, and coagulase tests, serology, DNA fingerprinting and phage typing. Some commercial rapid test kits are also available for the same. Antibiotics such as penicillin, cloxacillin, methicillin, vancomycin, oxacillin, cefotaxime, ceftriaxone, cephalosporin and rifampin are used for its treatment. Maintaining clean and hygienic conditions can be the control measures.

2.6.3.6 Syphilis

It is an STD transmitted by the spirochete *Treponema pallidum*. The pathogen is transmitted to the human body through mucous membranes or minor breaks or abrasions of the skin. It then enters the lymph nodes and then spread through the body. Symptoms include small, painless, reddened ulcer or **chancre** with a hard-rough boundary at the infection site [25]. Contact with the chancre during sexual intercourse may result in disease transmission. Syphilis is diagnosed by thorough physical examination, lumbar puncture (collection of spinal fluid to test pathogen) and dark-field microscopy of skin lesions and immunofluorescence examination of fluids taken from the lesions or sores. An antitreponemal antibody is formed in the host upon infection and hence the serological tests in case of syphilis are important.

The traditional non-treponemal tests such as **V**enereal **D**isease **R**esearch **L**aboratories test (VDRL), **R**apid **P**lasma **R**eagin test (RPR) and Wassermann test are done to identify non-treponemal antigens. Other treponemal-specific tests are performed to confirm the positive non-treponemal tests. These specific tests detect antibodies present in the host against antigenic components of *T. pallidum*. **E**nzyme **i**mmuno**a**ssay (EIA) for anti-treponemal IgG, **f**luorescent **t**reponemal ABS test (FTA-ABS), *T. pallidum* **i**mmobilization (TPI), *T. pallidum* complement fixation, *T. pallidum* **h**em**a**gglutination (TPHA) are the specific tests performed to find out treponemal antibodies [35]. Penicillin can be given to patients for syphilis treatment.

2.6.3.7 Tetanus

This is caused by *Clostridium tetani*, an anaerobic Gram-positive spore former whose endospores are present in the environment. They are transmitted *via* skin abrasions or wounds. Symptoms include uncontrolled stimulation of skeletal muscles surrounding the wound and jaw muscle tightening. With severity, the face muscles also go into spasms producing risus sardonicus. The person develops board like rigidity, **opisthotonos** (muscle spasms causing backward arching of the back, head, neck and spine). Tetanus toxoids are employed for prevention of the disease. Control for pathogen is not possible because of the wide propagation of the bacterium in the soil and the long survival of its endospores.

2.6.4 Foodborne and waterborne disease

Food- and water-mediated infections mainly cause gastroenteritis. Many microbes present in food and water can cause gastrointestinal infection such as abdominal pain, nausea, vomiting and diarrhea.

2.6.4.1 Botulism

It is caused by *Clostridium botulinum*, an obligate Gram-positive, rod-shaped, spore-forming anaerobic bacteria found in soil and aquatic environments. Their most important property is that they are spore formers. These endospores produce toxin during germination. This botulinum toxin is neurotoxic in nature and interferes with neural transmission by blocking acetylcholine release causing muscle paralysis. Acetylcholine is a principal neurotransmitter at neuromuscular junction. The muscles do not contract as a response, and other symptoms include vision impairment, difficulty in swallowing and speaking, weakening muscles, nausea and vomiting [55]. Hemagglutination tests are performed as a measure of laboratory diagnosis.

Antitoxins are generally administered to cure the disease. This disease is more common in infants.

2.6.4.2 Staphylococcal food poisoning

It is caused by *Staphylococcus aureus* which is present in inappropriately cooked or stored food. It is a Gram-positive coccus and also these bacteria are resistant to heat, drying and radiation. They produce enterotoxins (heat stable) in food; six of the toxins are designated as A, B, C1, C2, D and E. These toxins act as neurotoxins and possess superantigenic activity. Diagnosis is based on the culturing of the bacteria from foods. Treatment is done with fluid and electrolyte replacement. Food contamination should be avoided in order to control food poisoning.

2.6.4.3 Traveler's diarrhea

This is a gastrointestinal disorder which results when a traveler gets infected with specific virus, bacteria or fungi usually absent in their home environment. One of the most common bacteria causing diarrhea is *E. coli*. This bacterium circulates in the resident population, typically without causing symptoms due to the immunity afforded by previous exposure to the pathogen [45]. Other causal agents of the disease are *Campylobacter jejuni*, *Shigella* and *Salmonella*.

2.6.4.4 Typhoid fever

Typhoid fever is caused by *Salmonella typhi* and is developed by ingestion of food or water contaminated by feces of infected humans or animals. Symptoms include fever that can be high as 103–104 °F, stomach pain, headache, diarrhea or constipation, anorexia, malaise, cough and loss of appetite which last for several weeks. The Widal test is done to detect typhoid fever in patients. Ceftriaxone, trimethoprim–sulfamethoxazole or ampicillin are used for treating the fever.

2.6.5 Sepsis and septic shock

Sepsis is defined as the condition resulting from the presence of bacteria or their toxins in blood or tissues. Whereas **septicemia** is the clinical name for blood poisoning. In this condition, the pathogens or their toxins are found in the blood. **Septic shock** is the sepsis with hypotension (dangerously low blood pressure due to shock) along with abnormal cellular metabolism and other abnormalities. This kind of shock is the most common cause of death in intensive care units. Sepsis is the systemic response to the microbial infection [12]. Gram-positive bacteria, fungi and endotoxin-containing Gram-negative bacteria can cause sepsis leading to septic shock. Gram-negative organisms causing sepsis are *Escherichia coli*, *Klebsiella* spp., *Enterobacter* spp. and *Pseudomonas aeruginosa*. The Gram-negative bacteria contain LPS in their outer membrane which is the main reason behind the septic shock caused by the bacterium. The pathogens may directly enter the circulatory system and release toxins (endotoxins, teichoic acid antigen and exotoxins). These toxins cause the secretion of endogenous mediator from endothelial cells, plasma cells (monocytes, macrophages and neutrophils) and plasma cell precursors. These mediators have effects on heart, and other body organs leading to abnormal blood clotting, kidney failure, stroke and liver failure. Patient is either recovered or the septic shock leads to the death because of multiple organ failure.

2.6.6 Dental infections

The dental pathogens are called as **odontopathogens**, leading to tooth decay and dental diseases.

2.6.6.1 Dental plaque

Saliva is a natural defense against the bacterial infection. Dental plaque is formed by pathogens *Streptococcus gordonii*, *S. oralis* and *S. mitis*. After they attach to the enamel, several other opportunistic pathogens also arrive causing infection [31]. *Actinomyces viscosus*, *A. naeslundii* and *S. gordonii* are the later pathogens to attack. *Streptococcus mutans* and *S. sobrinus* are also the causative agents of plaque. These streptococcal pathogens produce glucosyltransferase, which polymerizes the glucose into extracellular water-soluble and water-insoluble glucan polymers and other polysaccharides. **Glucans** are branched chain polysaccharides composed of glucose units; they act as a binding material for the bacterial cells forming plaque.

2.6.6.2 Dental caries

This is a bacterial infection causing demineralization of the tooth by acids formed by bacteria. Acids are formed by bacterial fermentation of food debris deposited in the mouth. A whitish lesion is formed after demineralization which turns brown afterward forming a cavity. Once the pathogen reaches the hard enamel, bacteria can then invade the dentin and pulp of the tooth and cause its death. No drugs are available to prevent dental caries.

References

[1]   AlonsoDe VE, Verheul AFM, Verhoef J, Snippe H, *Streptococcus pneumoniae*: virulence factors, pathogenesis, and vaccines, Microbiol Rev, 1995, 59(4), 591–603.
[2]   Aral SO, Holmes KK, Sexually transmitted diseases in the AIDS era, Sci Am, 1991, 264(2), 62–69.
[3]   Avery OT, MacLeod CM, McCarty M, Studies on the chemical nature of the substance inducing transformation of pneumácoccal types, Induction of transformation by a deoxyribonucleic acid fraction isolated from Pneumococcus type III, J Exp Med, 1944, 79(2), 137–58.
[4]   Barbour AG, Laboratory aspects of Lyme borreliosis, Clin Microbiol Rev, 1988, 1(4), 399–414.
[5]   Bayles KW, The bactericidal action of penicillin: new clues to an unsolved mystery, Trends Microbiol, 2000, 8(6), 274–78.
[6]   Berg CM, Berg DE, Jumping genes: the transposable DNAs of bacteria, Am Biol Teach, 1984, 46(8), 431–39.
[7]   Bottcher HM, Wonder Drugs: a History of Antibiotics, Philadelphia, J. B. Lippincott, 1964.
[8]   Buchanan RL, Cygnarowicz ML, A mathematical approach toward defining and calculating the duration of the lag phase, Food Microbiol, 1990, 7(3), 237–40.
[9]   Cassady-Cain RL, Blackburn EA, Alsarraf H, et al., Biophysical Characterization and Activity of Lymphostatin, a Multifunctional Virulence Factor of Attaching and Effacing *Escherichia coli*, J Biol Chem, 2016, 291(11), 5803–16.
[10] Cassatella MA, The production of cytokines by polymorphonuclear neutrophils, Immunol Today, 1995, 16, 21–26.
[11] Clemens DL, Characterization of the *Mycobacterium tuberculosis* phagosome, Trends Microbiol, 1996, 4(3), 113–18.
[12] De Backer D, Dorman T, Surviving sepsis guidelines: a continuous move toward better care of patients with sepsis, JAMA, 2017, 317(8), 807–08.
[13] Dressler D, Potter H, Molecular mechanisms in genetic recombination, Annu Rev Biochem, 1982(51), 727–61.
[14] Falkow S, What is a pathogen?, ASM News, 1997, 63(7), 359–65.
[15] Flannagan RS, Cosio G, Grinstein S, Antimicrobial mechanisms of phagocytes and bacterial evasion strategies, Nat Rev Microbiol, 2009, 7, 355–366.
[16] Franklin TJ, Biochemistry and Molecular Biology of Antimicrobial Drug Action, New York, Chapman & Hall, 2001.
[17] Friedman RL, Pertussis: the disease and new diagnostic methods, Clin Microbiol Rev, 1988, 1(4), 365–76.
[18] Gill DM, Bacterial toxins: a table of lethal amounts, Microbiol Rev, 1982, 46, 86–88.
[19] Gootz TD, Discovery and development of new antimicrobial agents, Clin Microbiol Rev, 1990, 3(1), 13–31.
[20] Groisman EA, Ochman H, Pathogenicity islands: bacterial evolution in quantum leaps, Cell, 1996, 87, 791–94.
[21] Hare R, The Birth of Penicillin, Atlantic Highlands, NJ, Allen and Unwin, 1970.
[22] Henderson B, Oyston P (Eds.). Bacterial Evasion of Host Immune Responses (Advances in Molecular and Cellular Microbiology). Cambridge: Cambridge University Press, 2003.
[23] Henning U, Determination of cell shape in bacteria, Annu Rev Microbiol, 1975, 29, 45–60.
[24] Hilchie AL, Wuerth K, Hancock RW, Immune modulation by multifaceted cationic host defense (antimicrobial) peptides, Nat Chem Biol, 2013, 9, 761–68.
[25] Hook EW, Marra CM, Acquired syphilis in adults, N Engl J Med, 1992, 326(16), 1060–69.
[26] Hume DA, Macrophages as APC and the dendritic cell myth, J Immunol, 2008, 181, 5829–5835.

[27] Huycke MM, Sahm DF, Gilmore MS, Multiple-drug resistant enterococci: the nature of the problem and an agenda for the future, Emerg Infect Dis, 1998, 4(2), 239–49.

[28] Ippen-Ihler KA, Minkley EG Jr., The conjugation system of F, the fertility factor of *Escherichia coli*, Annu Rev Genet, 1986, 20, 593–624.

[29] Iseman M, Treatment of multidrug-resistant tuberculosis, N Engl J Med, 1993, 329(11), 784–91.

[30] Isenberg HD, Pathogenicity and virulence: another view, Clin Microbiol Rev, 1988, 1(1), 40–53.

[31] Jenkinson HF, Adherence and accumulation of oral streptococci, Trends Microbiol, 1994, 2(6), 209–12.

[32] Koch AL, What size should a bacterium be? A question of scale, Annu Rev Microbiol, 1996, 50, 317–48.

[33] Kolter R, Siegele DA, Tormo A, The stationary phase of the bacterial life cycle, Annu Rev Microbiol, 1993, 47, 855–74.

[34] Kotb M, Bacterial pyrogenic exotoxins as superantigens, Clin Microbiol Rev, 1995, 8(3), 411–26.

[35] Larsen SA, Steiner BM, Rudolph AH, Laboratory diagnosis and interpretation of tests for syphilis, Clin Microbiol Rev, 1995, 8(1), 1–21.

[36] Lederberg J, Tatum EL, Gene recombination in Escherichia coli, Nature, 1946, 158, 558.

[37] Lewis K, Hooper DC, Ouellette M, Multidrug resistance pumps provide broad defense, ASM News, 1997, 63(11), 605–10.

[38] Machado Paulo RL, Araújo Maria IAS, Carvalho L, Carvalho EM, Immune response mechanisms to infections, An Bras Dermatol, 2004, 79(6), 647–62.

[39] Mantovani A, Cassatella MA, Costantini C, Jaillon S, Neutrophils in the activation and regulation of innate and adaptive immunity, Nat Rev Immunol, 2011, 11, 519–31.

[40] Masters M, Transduction: host DNA transfer by bacteriophages, Encyclopedia of Microbiology, 2nd, San Diego, CA, USA, Academic Press, 2000, Vol. 4, 637–50.

[41] McEvedy C, The bubonic plague, Sci Am, 1988, 258(2), 118–23.

[42] Meselson M, Stahl FW, The Replication of DNA in Escherichia Coli, Proc Natl Acad Sci, 1958, 44, 671–82.

[43] Moura De SJ, Balbontín R, Durão P, Gordo I, Multidrug-resistant bacteria compensate for the epistasis between resistances, PLoS Biol, 2017, 15(4), 1741.

[44] Nassif X, So M, Interaction of pathogenic *Neisseriae* with nonphagocytic cells, Clin Microbiol Rev, 1995, 8(3), 376–88.

[45] Nataro JP, Kaper JB, Diarrheagenic *Escherichia coli*, Clin Microbiol Rev, 1998, 11(1), 142–201.

[46] Nathan C, Neutrophils and immunity: challenges and opportunities, Nat Rev Immunol, 2006, 6, 173–82.

[47] Olivier D, Route of transmission of *Staphylococcus aureus*, Lancet Infect Dis, 2017, 17(2), 124–25.

[48] Perry RD, Fetherston JD, *Yersinia pestis* – etiologic agent of plague, Clin Microbiol Rev, 1997, 10(1), 35–66.

[49] Pluddemann A, Mukhopadhyay S, Gordon S, Innate immunity to intracellular pathogens: macrophage receptors and responses to microbial entry, Immunol Rev, 2011, 240, 11–24.

[50] Prosser JI, Tough AJ, Growth mechanisms and growth kinetics of filamentous microorganisms, Crit Rev Biotechnol, 1991, 10(4), 253–74.

[51] Quagliarello V, Scheld W, Bacterial meningitis: pathogenesis, pathophysiology, and progress, N Engl J Med, 1992, 327(12), 864–71.

[52] Rasooly A, Rasooly RS, How rolling circle plasmids control their copy number, Trends Microbiol, 1997, 5(11), 40–46.

[53] Rietschel ET, Brade H, Bacterial endotoxins, Sci Am, 1992, 267(2), 54–61.

[54] Rogers HJ, Bacterial Cell Structure, Washington, American Society for Microbiology, 1983.

[55] Schantz EJ, Johnson EA, Properties and use of botulinum toxin and other microbial neurotoxins in medicine, Microbiol Rev, 1992, 56(1), 80–99.

[56] Silva MT, Correia-Neves M, Neutrophils and macrophages: the main partners of phagocyte cell systems, Front Immunol, 2012, 4(3), 174.

[57] Stewart GJ, Carlson CA, The biology of natural transformation, Annu Rev Microbiol, 1986, 40, 211–35.

[58] Todd JK, Toxic shock syndrome, Clin Microbiol Rev, 1988, 1(4), 432–46.

[59] Toro E, Shapiro L, Bacterial Chromosome Organization and Segregation, Cold Spring Harb Perspect Biol, 2010, 2(2), 349.

[60] Young KD, Bacterial morphology: why have different shapes?, Curr Opin Microbiol, 2007, 10(6), 596–600.

[61] Zughaier SM, Shafer WM, Stephens DS, Antimicrobial peptides and endotoxin inhibit cytokine and nitric oxide release but amplify respiratory burst response in human and murine macrophages, Cell Microbiol, 2005, 7, 1251–1262.

Malay Kumar Sannigrahi, Deepika
Chapter 3
Virology

Abstract: Virology includes aspects related to viruses. Viruses are obligate intracellular parasites. They are acellular infectious agents that cannot multiply without a living host. Viruses are responsible for a large number of diseases including some fatal disease for which it is still difficult to find a stable treatment. New viruses are also emerging continuously, presenting challenges for the scientists worldwide. In this chapter, viral classification, diseases, immune response, genetic mutations, control and diagnosis of viruses have been presented.

Keywords: viruses, classification, structure, pathogenesis, treatment

3.1 Introduction

Viruses are known to be obligate intracellular parasites. They are acellular infectious agents that cannot multiply without a living host. When found outside of host cells, they are metabolically inert particles. These particles have a protein coat surrounding them called capsid which may or may not be enclosed in a membrane. This protein coat (capsid) envelopes either deoxyribonucleic acids (DNA) or ribose nucleic acids (RNA), which codes for the virus proteins. When it meets a host cell, a virus can inject its genetic material into the host cell, hijacking the host's fundamental functions for its own replication and multiplication. Unlike most living things, viruses can only divide in the infected host cell. Viruses spread in many ways and have distinct mechanism of infection, pathogenesis and symptoms. Most of the time, they are usually eliminated by the host immune system, and diseases are prominent in immunocompromised host. This chapter discusses the mode of infection, life cycle, pathogenesis, diagnosis, control and prevention of various viral infections.

Malay Kumar Sannigrahi, Post Graduate Institute of Medical Education and Research, Chandigarh, India, e-mail: malaysannigrahi@gmail.com
Deepika, Thapar Institute of Engineering and Technology, Patiala, Punjab, India, e-mail: deepika0184@gmail.com

https://doi.org/10.1515/9783110517736-003

3.2 Viral classification, structure and multiplication

The classification of viruses involves naming and putting them into different taxonomic categories. However, viruses do not have ribosomes (therefore, ribosomal RNA is also absent, which is widely used in classification of other microorganisms); hence, they cannot be classified as per the three-domain classification scheme. There are two main classification systems adopted for viruses: The International Committee on Taxonomy of Viruses (ICTV) system and Baltimore classification system.

The International Union of Microbiological Societies selected ICTV which started in the 1970s for the nomenclature and classification of the viruses. Since then, they are developing, redefining, naming and maintaining universal virus taxonomy. The ICTV divided viruses into order (-virales), family (-viridae), subfamily (-virinae), genus (-virus) and species. The first report was published in 1971 with 43 virus groups. Recently, the Ninth Report was published with 1,327 pages and described 87 viral family totaling 2,284 species [1]. The major orders that span viruses with varying host ranges are the Herpesvirales, Caudovirales, Nidovirales, Ligamenvirales, Picornavirales, Mononegavirales and Tymovirales. They also created nine unassigned genera including Aumaivirus, Papanivirus, Sinaivirus, Virtovirus, Anphevirus, Arlivirus, Chengtivirus, Crustavirus and Wastrivirus in the order Mononegavirales, and eight new families including Botybirnaviridae, Genomoviridae, Lavidaviridae, Mymonaviridae, Pleolipoviridae, Pneumoviridae, Sarthroviridae and Sunviridae [1].

David Baltimore developed a system called the Baltimore Classification which groups or categorizes the viruses into families, based on their genetic material (like DNA and RNA) and their technique of replication [1]. The following classification of the viruses was proposed:
(I) double-stranded (ds) DNA viruses (e.g., adenoviruses, herpesviruses and poxviruses),
(II) single-stranded (ss) DNA viruses (sense strand) DNA (e.g., parvoviruses),
(III) ds RNA viruses (e.g., reoviruses),
(IV) (positive) ss (+ss) RNA viruses (sense strand) RNA (e.g., picornaviruses and togaviruses),
(V) (negative) ss (–ss)RNA viruses (antisense strand) RNA (e.g., orthomyxoviruses and rhabdoviruses),
(VI) ss RNA retroviruses (sense strand) RNA with DNA intermediate in life cycle (e.g., retroviruses) and
(VII) ds DNA retroviruses DNA with the RNA intermediate in life cycle (e. g., hepadnaviruses).

3.3 Heredity and variation in viruses

Genetic variation allows natural selection in viruses resulting in their continuous adaptation to a varying environment. Viruses face consistent environmental stress while passing from host to host and therefore keep on changing because of genetic selection. Moreover, they also undergo mutation and recombination leading to many genetic changes.

3.3.1 Mutations

Mutation results in a random genetic error leading to genetic variation in viruses. Viruses have excessive mutation rates. DNA viruses may undergo 10^{-8}–10^{-6} modifications per nucleotide during a single infection cycle [3]. The rate of mutation depends upon multiple processes, including intrinsic fidelity of the DNA/RNA polymerase that helps the virus to replicate, the replication mode of the virus, 3′-exonuclease activity of the viral enzyme, spontaneous nucleic acid damages in the virus during its infection cycle, post-replicative repair in the virus, various editing functions by host-encoded deaminases, any variation in nucleotide pools of the host, context template sequences of the virus based on its GC content and template structure of the virus [4]. Therefore, both the host and viral infection process leads to mutations in virus.

The RNA viruses have inherently high rate of mutations as compared to the DNA viruses because RNA polymerases, which are responsible for replicating the viral genome, do not contain proof reading and error editing functions that are present in cellular DNA polymerases [5]. RNA viruses may undergo the 10^{-6}–10^{-4} modifications per nucleotide during a single infection cycle [3]. Host-encoded factors like some DNA/RNA editing enzymes also regulate mutations in RNA viruses like Apolipo protein B mRNA-editing catalytic polypeptide-like enzymes, a cellular cytidine deaminase known to induce mutation in RNA viruses [6]. Comparing the genome size and rate of mutation, a direct correlation was observed, suggesting larger is the size of DNA/RNA, and more is the number of mutations [7].

3.3.2 Recombination

Recombination is another method of viral genetic variation. It usually happens among the members of the same virus types. Recombination occurs when viruses of two completely different viral strains infect a common host cell (coinfection), and their interaction during replication may generate virus progeny which will have the genes from the two different viruses (or both the parents); for example, α-herpesviruses and Western equine encephalitis (WEE) virus.

There are two major mechanisms of recombination: independent assortment and incomplete linkage.

An independent assortment was observed with viruses having segmented genomes during replication. In this process of recombination, new antigenic determinants are produced which results in viruses with new host range. Based on where the parental strands are from, it is divided into two types: when the recombining strands are of the same genetic unit, it is called intra-genomic recombination; and when the strands have different origin, it is known as inter-genomic recombination. The latter is often related to the regulation of gene expression [8]. Intra-genomic recombination occurs in DNA viruses by means of a break–rejoin mechanism. Inter-genomic recombination occurs in the closely related members HSV-1 and HSV-2 (75% similarity) belonging to human α-herpesviruses [9].

RNA viruses including retroviruses, picornaviruses and coronaviruses also show recombination [10]. RNA–RNA recombination can be a very precise event so that the crossover events between two recombining RNA sequences cannot be easily identified [11]. In general, (+) ssRNA viruses show high recombination rates than (–) ssRNA [12]. However, certain RNA viruses like Flaviviruses rarely show recombination, whereas reassortments are observed in Orthomyxoviridae such as influenza A virus [13, 14].

Recombination can also be used by various viruses to survive stress conditions. Bacteriophages, including phage T4, undergo multiplicity reactivation in which homologous recombinational repair on lethal genomic damage by pairing with each other and to produce a progeny [15].

3.4 Immune response to viral infections

Type I and III interferon (IFNs) inductions are the first line of defense, limiting viral replication and spread. There are specific epitopes of viruses responsible for immune response called pattern recognition receptors (PRRs). PRRs including C-type lectin receptors expressed by dendritic cells are crucial for generating immune responses to pathogens. Various other receptors are also expressed by most other cells like toll-like receptors (TLRs), intracellular retinoic acid-inducible gene-I (RIG-I)-like receptors (RLRs), nucleotide oligomerization and binding domain–like receptors and the Pyrin-HIN domain receptors [16, 20]. Viral infection activates these TLRs and RLRs, resulting in the IFN response.

When the host senses viral genetic material and/or viral oncoproteins leading to metabolic and cellular damage, it induces IFN response [16]. Viral DNA/RNA activates various signaling pathways in cell-like IFN-regulatory factor-3, NF-kappa B, dsRNA-dependent protein kinase R and/or the JNK2 pathway [17]. For example, treatment with IFNs induces apoptosis in various virally infected cells like herpesvirus-associated cancers of the lung, skin and brain [18–22]. Viral DNA/RNA can be detected by other receptors like RLR family, RIG-I and Mda5 (melanoma differentiation-associated gene 5) [23]. Natural killer cell is also an important immune effector cell that helps in controlling the virus infection and induces tissue damage during chronic viral infection by various mechanisms [24, 25].

3.5 Control of viral diseases

Viruses show huge variations in their epidemiology and pathogenesis which makes them difficult to control. Viral diseases can be mostly curbed by decreasing the exposure to virus. This can be done by (a) eliminating intermediate nonhuman reservoirs which allow the virus to stay alive, (b) eliminating the vector which transfers the virus to host and (c) improving sanitation to clean any viral contamination. Proper testing can be done to reduce viral infected samples like hepatitis B infection, which can be identified in the blood by detecting its surface antigen (Ag). Similar tests are used in blood bank for detecting antibodies to human immunodeficiency virus (HIV)-1, HIV-2, HTLV-I (human T-cell leukemia virus) and hepatitis C. This can also help identify and discard any contaminated blood units. Another approach is preventing viruses that are transmitted through body fluids by proper protection (like condoms against HIV and N-90/95 masks to avert coronavirus).

The best way to protect against viral disease is by immunization, i.e., using vaccines or antibodies produced against viral proteins (present in plasma of individuals recovered from viral illness). There are two types of immunization methods: active or passive. Active immunization is attained through using a noninfectious virus preparation to stimulate the body's immune system to produce its own antibodies, also known as vaccination. There are three types of viral vaccines: (1) **attenuated live viruses** (MMR combined vaccine for measles, mumps, rubella viral infection, vaccine available for rotavirus, smallpox, chickenpox, yellow fever also) which are nonpathogenic live viral preparations, when used can infect the recipient and can replicate/reproduce to generate a protective immune response without causing any disease. They confer lifelong immunity after one immunization dose. However, an important drawback is since the infection is with a live virus, there is always a chance that the virus may return to its pathogenic form and multiply in the body; (2) **killed virus preparations** are made by inactivating the virus or some component(s) of the virus using chemical or physical means. These kinds of vaccine often do not cause infection, but they are less effective also. This approach has been against polio, rabies, flu and coronavirus. (3) **Recombinant produced Ags** (viral protein subunit, recombinant-viral particles, polysaccharide and conjugate vaccines) use recombinant DNA strategy by cloning and expressing a viral surface gene into a suitable vector. This approach has been against hepatitis B virus (HBV).

mRNA vaccines are a new type of vaccine developed recently to protect against infectious diseases like COVID-19 mRNA vaccines. They make viral proteins to trigger an immune response and have no risk of causing disease.

Passive immunity involves the use of passive vaccine like the plasma of HBV or coronavirus-infected individual which is used to administer antibodies against HBV or coronavirus formed in another host. This process uses human immune-globulins from a recovered individual which are used to guard sick people who have been recently exposed to a disease and need imminent help to survive because of their complications.

Antiviral agents are chemical agents that prevent viral replication inside the cell. They inhibit one or more than one step in the life cycle of the virus. However, these antivirals are often toxic to various degrees to the normal cells in the human body and have a restricted spectrum of their activity. They are of three main types: virucidal agents, virustatic agents and immunomodulators. Virucidal agents are those that can inactivate intact viruses, e.g., podophyllin used for wart treatment. Virustatic agents do not kill the virus but inhibit replication of the virus. They can inhibit the fusion of virus with the host cell like maraviroc (CCR5 antagonist) and enfuvirtide; inhibit replication of the virus, e.g., efavirenz and nevirapine; inhibit protein synthesis of the virus, e.g., darunavir, atazanavir and indinavir for HIV, boceprevir for HCV. Immunomodulators help in clearing the virus, e.g., imiquimod (Aldara) for human papillomavirus (HPV)-infected genital warts, and alitretinoin gel, for the treatment of cutaneous viral diseases [26, 27].

3.6 Laboratory diagnosis of virus infection and antivirus therapy

Clinical testing of viral diseases involves interpreting the DNA/RNA or its corresponding antibodies or its activity which in turn can determine the presence or absence of the active viral infection. The practicality and dependability of laboratory test results largely depend on the sensitivity and specificity of these tests.

Virological testing is done in a wide variety of samples. The type of sample depends upon the category of viral infection which has to be diagnosed. Blood or body fluid is used for serological testing for antibody and Ag detection in a variety of different infections. For respiratory viruses (like influenza), sputum, gargles and bronchial washings are used. Fecal samples are generally used to identify the existence of viral gastroenteritis agents. Other samples include cerebrospinal fluid, biopsies and tissues, and dried blood spots.

Complement fixation test is one of the widely used methods for testing viral infection. It is a convenient and easy technique, which requires inexpensive material. Viral surface has antigenic regions that can bind to specific antibodies. The complement system nonspecifically attaches with the antigen–antibody complex. Thus, in the presence of the complex attached to the complement, no free complement is available to react with sensitized sheep red blood cells (RBCs used as indicators), and which remain unlysed. However, it is labor-intensive and lacks sensitivity.

Hemagglutination inhibition test is another widely used viral detection test. It relies on the capacity of a viral envelope protein named hemagglutinin (HA); to bind to erythrocytes (RBC) and to form a lattice pattern termed "agglutination." However, the assay has limitations for its sensitivity is dependent on concentration of the virus. In some cases, at lower viral concentration, they may bind to the RBCs, thus preventing them from settling out of suspension giving false-negative results.

Various serological procedures can be used to detect specific viral Ag in the blood or body fluids, viral nucleic acid, antiviral antibodies, etc., in the blood or body cells of patient. Cytological or histological examination may be used for detection of the presence of a cell-mediated immune response of cells from the site of the infection where a characteristic viral cytopathic effect (CPE) is produced.

Cell culture is used for isolating viruses, e.g., rhesus monkey kidney cells are used for isolation of influenza A virus. Growth of virus can be checked by CPE, which involves the death of viral infected cells. However, this method cannot be used for rapid diagnosis as the time required for isolating viruses ranges from weeks to months. Moreover, highly skilled, experienced personnel and proper biosafety facilities are required for handling highly contagious viruses.

One of the tests for definitive identification of a virus is performed using immunofluorescence (IF) staining using antibodies against specific viral antibodies followed by high-resolution microscopy for direct visualization of the viral particles. For example, diagnosis of hepatitis virus infection using HBsAg (hepatitis B viral surface Ag) or HBeAg (hepatitis B viral e Ag) or for HIV infection, p24 viral Ag can be detected. Direct microscopic examination of clinical specimens such as biopsy materials or skin lesions along with staining with certain dyes helps in detecting viral infection. High-resolution electron microscopy (EM) can be used to directly visualize and count the virus particles in body fluids, stools or histopathologic samples.

Besides these methods, PCR (polymerase chain reaction) can be used to detect viral nucleic acids (DNA/RNA) which uses specific primer/probe for detection of the specific gene sequence of that virus. Other methods include nucleic acid amplification tests to determine the viral load (for HBV, HCV, HIV, influenza viruses, etc.) which uses quantitative real-time PCR. This is now considered a reference or "gold standard" method for viral detection [28].

Following are the example of virus and short description of their mode of infection, life cycle, pathogenesis, detection, control and treatment:

3.7 Respiratory infective myxoviruses

The respiratory infective myxoviruses may be broadly divided into two subgroups: the influenza subgroup (influenza virus types A, B and C) and the non-influenza subgroup (parainfluenza types 1 (HA-2), 2 (CA), 3 (HA-1) and 4, mumps, measles and respiratory syncytial viruses).

3.7.1 Type-I myxovirus

Orthomyxoviruses include influenza virus types A, B and C. These are spherical or filamentous viruses in which the viral envelope has HA protein as projected spikes and neuraminidase in between these spikes [29].

3.7.1.1 Mode of infection

These are pneumotropic viruses that multiply only in the epithelial cells of the respiratory tract. These viruses are inhaled and penetrate the outer epithelial layer of the respiratory tract.

3.7.1.2 Life cycle

The virus surface receptor binds to a sialic acid-containing receptor. The viral envelope fuses with the endosomal membrane, its genome escapes degradation and viral ribonucleoproteins (vRNPs) enter the nucleus where transcription and replication of the viral genome occur. Then the vRNPs are exported from the nucleus; viral assembly happens in the cytosol and budding occurs at the plasma membrane. These viruses remain viable in a dried state and at low temperature for a very long period [30].

3.7.1.3 Pathogenesis

The virus multiplies in the epithelial cells of the trachea, bronchi, bronchiole and alveolar regions. This results in loss of cilia in the ciliated epithelium. This is followed by various opportunistic infections of secondary bacterial flora (streptococci, pneumococci, *Haemophilus influenzae*, etc.) that give rise to many complications (pneumonia, encephalitis, influenzal meningitis and otitis media).

3.7.1.4 Detection

Diagnostic tests include serological testing, rapid Ag testing, quantitative reverse transcription PCR (qRT-PCR), IF assays with high-resolution microscopy and rapid molecular assays.

3.7.1.5 Control

Influenza virus is spread by the air droplet route by sneezing, coughing and talking. Human-to-human transmission is a common route. Transmission can be prevented by isolating the patient [31].

3.7.1.6 Treatment

Infection is highly contagious; it is known to cause epidemics and pandemics. Monovalent or polyvalent vaccines are available. Antiviral agents like L-amantadine hydrochloride can block the entry of influenza virus and prevent viral replication.

3.7.2 Type-II myxovirus

Paramyxoviruses include mumps, measles, canine distemper and rinderpest. They have large genomes (>15,000 bases), pleomorphic and can be subdivided into para-influenza virus, paramyxovirus proper, measles virus (Morbillivirus), mumps virus (Rubulavirus) and pneumoviruses such as respiratory syncytial virus.

3.7.2.1 Mode of infection

These viruses also spread through the air droplet route.

3.7.2.2 Life cycle

Virus adheres to the host cell and replicates in the cytoplasm. They have nonsegmented RNA and follow negative-stranded RNA virus replication model, except in mumps viruses, where a single gene produces several variant mRNAs [32].

3.7.2.3 Pathogenesis

Often results in upper respiratory tract lesions (rhinitis, pharyngitis and bronchitis) in children and produces mild disease in adults. Mumps is a mild benign "childhood" disease that causes mild inflammation of glandular tissue in the head and neck. The respiratory syncytial virus results in cold-like infection of adult's nasopharynx. Diseases like measles and the closely related distemper and rinderpest viruses are serious and often fatal. Canine distemper infections cause high mortalities in domestic, wild and marine animals.

3.7.2.4 Detection

qRT-PCR assays, IF assays with high-resolution microscopy and rapid molecular assays are available.

3.7.2.5 Control

Heating and disinfection result in killing. However, once exposed, long-term immunity develops.

3.7.2.6 Treatment

Vaccines are usually available as monovalent or polyvalent forms [33].

3.7.3 Gastrointestinal virus and rhinovirus

Viral gastroenteritis results in inflammation of the lining of the stomach, small intestine and large intestine [34]. This is followed by vomiting and diarrhea leading to dehydration. They are highly contagious and extremely common. Viral gastroenteritis is caused by four types of viruses described below.

3.7.3.1 Life cycle

They are positive-stranded RNA viruses and does not have an outer membrane. They have icosahedral capsid which is 30 nm and packages a 7.2–7.5 kb genome. The four capsid proteins VP1, VP2, VP3 and VP4 form the outer coat. They are formed by 12 pentamers, each of which has 5 protomer subunits. Moreover, translation of viral protein requires an internal ribosomal entry site (IRES) in the 5′-untranslated region of the genome. Adenovirus are medium-sized (90–100 nm) dsDNA viruses encoding 22–40 genes that are between 26 and 48 kbp. In case of rotaviruses and caliciviruses, the nonstructural proteins are encoded in the 5′-end of the genomic RNA and the structural proteins in the 3′-end. Virus uptake is mediated either by endocytosis process which could be clathrin-dependent or -independent. When the virus is endocytosed, it undergoes a conformational change due to either drop in pH or by receptor binding. The outside hydrophobic domains are opened, and the viral genome is released in the cytoplasm.

3.7.3.2 Pathogenesis

3.7.3.2.1 Rotavirus
Rotavirus primarily causes gastroenteritis among infants aged 3 to 15 months old and young children. Symptoms usually appear 1 to 3 days after exposure. Their infections result in diarrhea along with fever and abdominal pain, and often early childhood death worldwide. They can also infect adults but the symptoms are milder [35].

3.7.3.2.2 Caliciviruses
The Caliciviridae on negative stain EM appears to have a distinctive cup-shaped depression on their surface, giving them their name. There are currently seven species in this family, divided among five genera and can infect all organisms. They are known to cause respiratory diseases (feline calicivirus), hemorrhages (rabbit hemorrhagic disease virus) and gastroenteritis (Norwalk viruses). Norovirus causes viral gastroenteritis, often 1–2 days after exposure in which typically adults experience nausea, vomiting, diarrhea, abdominal cramps, fatigue, headache and muscle aches [36, 37].

3.7.3.2.3 Adenovirus
Adenoviruses infect children younger than 2 years old causing conjunctivitis and tonsillitis. The virus has 52 serotypes, and some strains affect the gastrointestinal tract, causing vomiting and diarrhea. It can be transmitted through a fecal–oral route, aerosolized droplets or exposure to the infected tissue or blood. Symptoms typically appear 8–10 days after exposure. The upper respiratory tract infection by adenovirus occurs frequently. Sometimes they lead to conjunctivitis, tonsillitis (which may look exactly like strep throat), an ear infection, bronchiolitis or pneumonia in children, both of which can be severe [38, 39].

3.7.3.2.4 Astrovirus

Human astroviruses (HAstVs) may be classified into three groups: the classic (MAstV 1), HAstV-MLB (MAstV 6) and HAstV-VA/HMO (MAstV 8 and MAstV 9) groups. The classic HAstVs contain eight serotypes and this is the reason for around 9% of all acute nonbacterial gastroenteritis in children and adults with compromised immune system. Viral particles are often found in fecal matter and in epithelial intestinal cells in these cases. Symptoms usually appear 3–4 days after exposure but are milder than the symptoms of norovirus or rotavirus infections. The main symptoms are watery diarrhea followed by nausea, vomiting, fever, malaise and abdominal pain. The absorptive function of the intestine is disrupted, resulting in rapid loss of fluids and electrolytes and loss of intestinal epithelial barrier permeability [40].

3.7.3.3 Detection

Gastroenteritis can usually be diagnosed by a thorough medical examination, and laboratory test like "rapid stool test" can help detect rotavirus or norovirus infection. Rhinovirus was detected by PCR. Severe dehydration and low blood count can reveal electrolyte imbalances.

3.7.3.4 Control

Most cases resolve over time without specific treatment within 10–12 days. Transmission is mainly by aerosol and direct body contact, which can be interrupted by using mask and quarantining the infected individual.

3.7.3.5 Treatment

Symptomatic treatment can be done by over-the-counter drugs such as loperamide (Imodium) and bismuth subsalicylate (Pepto-Bismol). Live oral vaccine approved by the FDA (USA) is available which include pentavalent RotaTeq vaccine and Rotarix. RotaTeq is used for children in three doses at 2, 4 and 6 months of age after birth. Rotarix is given twice, firstly at 6 weeks after birth and then at 24 weeks.

3.7.4 Hepatitis A virus

Hepatitis A virus (HAV) is a small, unenveloped symmetrical RNA virus. It is transmitted by the fecal–oral route and causes mild infections to epidemic hepatitis [41].

3.7.4.1 Life cycle

HAV is a positive-strand RNA virus. Its genome is approximately 7.5 kb in length, linked covalently to a small polypeptide (VPg) at its 5′-terminus and has a 3′-poly-A tail. The 5′-terminal nontranslated RNA segment also contains IRES and numerous highly ordered RNA structures.

After entry into the host cell, the viral RNA induces the host cell to produce the viral polyprotein without affecting the protein synthesis of the cell. The viral proteins include a protease for the synthesis of structural proteins and a polymerase for the replication of viral RNA. The virus is noncytopathic, i.e., does not cause lysis of the host cell when grown in cell culture. However, in vivo, it causes necrosis of parenchymal cells and histiocytic periportal inflammation [42].

3.7.4.2 Pathogenesis

HAV enters the body by ingestion of contaminated food or intestinal infection. It then spreads through the bloodstream, to the liver where it incubates for about 4 weeks. Before the onset of clinical symptoms like jaundice, large quantities of virus are shed in the feces. Main pathologic symptoms include conspicuous focal activation of sinusoidal lining cells; accumulations of lymphocytes and histiocytes in the parenchyma, cytolytic necrosis of hepatocytes in the periportal areas, and formation of acidophilic bodies and focal degeneration.

3.7.4.3 Detection

Virus particles are detectable in feces during the incubation period. Serologic tests including immune EM, complement fixation, immune adherence hemagglutination, radioimmunoassay and enzyme-linked immunoassay (EIA) are widely used for HAV detection. Immune adherence hemagglutination and solid-phase type of assays are sensitive and specific.

3.7.4.4 Control and treatment

The hepatitis A vaccine is a series of two vaccines and given to children between ages 12 and 18 months. For adults, Hep-A vaccine is combined with the hepatitis B vaccine. However, bed rest is often recommended in adults and if symptoms like vomiting or diarrhea occurs, oral hydration therapy is done.

3.7.5 Hepatitis B virus

HBV is a dsDNA virus belonging to the hepadnavirus group. It is most common in the human population; however, it infects other mammals also [43].

3.7.5.1 Life cycle

HBV is an enveloped virus. The virion is a 42-nm particle with 27 nm core. The virus has surface protein (HBsAg) embedded in membranous lipid derived from the host cell. It replicates in the liver by reverse transcription, however, can get integrate into the chromosomal DNA of hepatocytes. The 3.5-kb genomic transcript codes two different transcripts: a pregenomic RNA serves as the template for reverse transcription and a messenger RNA for core and polymerases.

3.7.5.2 Pathogenesis

If infected, most individuals have immune clearance of infected hepatocytes. Infection can spread through body fluid contact, especially blood-to-blood transmission may occur. In case of acute hepatitis, symptoms like signs of jaundice include tiredness, flu-like symptoms, dark urine, pale stool, abdominal pain, loss of appetite, unexplained weight loss and yellow skin and eyes might occur [44].

3.7.5.3 Detection

Virus infection in infected hepatocytes can result in HBV DNA, the S1 proteins (HBsAg) and HBeAg in serum. Laboratory diagnosis involves the presence of HBsAg or by detecting viral DNA. Virus particles can be observed by high-resolution microscopy [45].

3.7.5.4 Control and treatment

The CDC recommends hepatitis B vaccinations to all newborns. Generally, three vaccines are given during the first 6 months of childhood. Chronic conditions are treated with antiviral medications. IFNα is given once a week for 48 weeks. Nucleoside/nucleotide analogues (NAs) including adefovir (Hepsera), entecavir (Baraclude), lamivudine (Epivir-HBV, Heptovir, Heptodin), telbivudine (Tyzeka) and tenofovir (Viread) may be used [46].

3.7.6 Hepatitis C virus

Hepatitis C virus (HCV) is an enveloped ssRNA virus. It is related to flaviviruses; however, it has six genotypes and various subtypes with differing geographical distribution and is associated with chronic liver disease.

3.7.6.1 Life cycle

The virus attaches itself to the cell membrane, and protein core penetrates the plasma membrane by merging its lipid coat with the cell's outer membrane and enters the cell. The HCV genome consists of a single uninterrupted open-reading frame (ORF) flanked by nontranslated regions at both its 5′ and 3′ ends. The RNA has structured RNA elements with numerous stem loops and a pseudoknot. The single polyprotein precursor codes for approximately 3,000 amino acids, which are posttranslational modified by cellular (e.g., signal peptidases) and viral proteases (NS2 and NS3/4A) to produce 10 different viral proteins, which include core and envelope glycoproteins (gps), E1 and E2, p7, NS2, NS3, NS4A, NS4B, NS5A and NS5B. Once the capsomeres are formed, they self-assemble and fully encapsulate the RNA virus [47].

3.7.6.2 Pathogenesis

HCV has incubation period of 2 weeks to 6 months. After initial infection, many do not show any symptoms. Symptomatic cases may exhibit illness, tiredness, decreased appetite, vomiting, abdominal pain, dark urine, gray-colored feces, joint pain and jaundice. During incubation period, the surface Ag can be detected. When liver damage starts, viral DNA is observed in circulation followed by elevated serum transaminases. Antibodies against viral Ag develop in infected persons serum within 2–10 weeks and remain years after recovery. The infection clears in 15–45% of the individuals without any treatment. For those who develop chronic HCV infection, cirrhosis of the liver can occur in 15–30% cases within 20 years [48].

3.7.6.3 Detection

HCV infection is detected by two methods: serological test for anti-HCV antibodies (against conserved C22, NS3 and NS4 domains) identifies people who have been infected with the virus. This is followed by a nucleic acid test for HCV RNA for chronic infection. Recently developed tests included Ags from the nucleocapsid and further nonstructural regions of the genome which appear relatively early in infection [49].

3.7.6.4 Control and treatment

Many HCV patients do not need treatment as the infection clears by the immune system of the body. Regular checkups and blood tests to monitor liver function are perquisites for most infections. Antiviral medications such as sofosbuvir, daclatasvir and the sofosbuvir/ledipasvir combination may be prescribed to chronic cases for several weeks to rid their body of the virus.

3.7.7 Hepatitis D virus

Hepatitis D virus (HDV) is an RNA virus.

3.7.7.1 Life cycle

HDV is a defective RNA virus that requires HBV virus for its replication and assembly of new virions. Its genome has only one actively transcribed ORF which encodes for two isoforms of hepatitis delta Ag: small and large delta Ags (S-HDAg and L-HDAg). S-HDAg helps in the initiation of the viral genome replication, whereas L-HDAg is required for the assembly of new virion particles and further infection process [50].

3.7.7.2 Pathogenesis

HDV is transmitted through blood via intravenous drug use and multiple blood transfusions. It replicates independently in hepatocyte, but for dissemination it requires a helper virus HBsAg. Coinfected with HBV and HDV results in acute hepatitis caused by HBV, which may lead to direct cytotoxic effect or host-mediated immune response [51, 52].

3.7.7.3 Detection

HDV is confirmed by detection of HDV RNA by qRT-PCR in serum. However, no specific treatment strategy is available.

3.7.7.4 Control and treatment

HDV infection can be controlled by prevention of HBV transmission through immunization with available hepatitis B vaccines and safe blood transfusion. There is no specific treatment. Pegylated IFN-alpha and antiviral NAs can be used to treat symptoms [54].

3.7.8 Hepatitis E virus

Hepatitis E virus (HEV) is a nonenveloped, ssRNA virus. It causes acute hepatitis in the Asian subcontinent, Middle East and parts of Africa.

3.7.8.1 Life cycle

HEV belongs to the genus Hepevirus and family Hepeviridae. There is very limited knowledge on HEV pathobiology because of the absence of a reliable in vitro propagation system. The 7.2 kb genome has three ORFs with 5′- and 3′-*cis*-acting elements. HEV replicates in the cytoplasm, and the full genomic RNA encodes nonstructural proteins like methyl transferase, protease, helicase and replicase from ORF-1 and capsid protein from ORF-2. The virus uses host machinery to replicate through a negative-sense, replicative intermediate RNA. The viral capsid protein consists of 660 amino acids, which is produced from ORF2 is responsible for virion assembly and immunogenicity [50].

3.7.8.2 Pathogenesis

HEV1 and HEV2 spread in humans by the fecal–oral route through contaminated water. HEV3 and HEV4 are zoonotically transmitted from animal reservoirs. The incubation period after exposure is 2–10 weeks with early symptoms including mild fever, abdominal pain, vomiting, itching (without skin lesions) and skin rash. This is followed by similar symptoms to other hepatitis which include jaundice (yellow discoloration of the skin and sclera of the eyes), with dark urine and pale stools, and a slightly enlarged, tender liver (hepatomegaly) [51, 52].

3.7.8.3 Detection

Hepatitis E symptoms are not clinically distinguishable from other viral hepatitis. qRT-PCR may be used to detect the HEV RNA in blood and/or stool [53].

3.7.8.4 Control and treatment

Two candidate HEV vaccines are under development, first from 56 kDa protein encoded by ORF2 of an HEV1 strain and the second vaccine, HEV 239, is a 26 kDa protein encoded by ORF2. Infection spread can be prevented by limiting the use of undercooked meat, especially pork products. However, clean drinking water and improving the sanitary infrastructure can also prevent infection in poor developing countries [54–56].

3.7.9 The GB hepatitis viruses

The GB hepatitis viruses (GBV-A, GBV-B and GBV-C) are positive-stranded RNA viruses. They are closely related to each other and are not genotypes of HCV. Preliminary studies suggest the prevalence of GB viruses in 3–14% of patients. GBV-C RNA was observed in patients with chronic hepatitis, cryptogenic cirrhosis and primary hepatocellular carcinoma, where other involvements of other causal agents such as acute A to E viral hepatitis are eliminated [57].

3.8 Arthropod-borne and rodent-borne viral disease

Arbovirus is a virus transmitted by arthropod vectors. However, human–vector–human cycle may also occur, e.g., yellow fever and dengue [58].

In general, arboviruses belong to the following three families:

(a) Togaviruses, e.g., include Venezuelan equine encephalitis (VEE) virus, WEE virus and Eastern equine encephalitis (EEE) virus.

(b) Bunyaviruses, e.g., include Crimean-Congo hemorrhagic fever, Sicilian sandfly fever and Rift Valley fever.

(c) Flaviviruses, e.g., include dengue, Zika virus, yellow ever and encephalitis.

3.8.1 Togaviruses

The family Togaviridae contains two genera: Alphavirus and Rubivirus (rubella virus).

3.8.1.1 Alphavirus

They are small, enveloped (+)ssRNA virus. They replicate and spread by mosquito–vertebrate or human–mosquito cycles.

3.8.1.2 Life cycle

These enveloped viruses are spherical, 60–70 nm, ssRNA genome around 11.7 kb long. Its mRNA codes for nonstructural proteins. The viral envelope has two surface gps that help in the attachment, fusion and penetration of the virus in the host cells. Virions mature by budding through the plasma membrane [59].

3.8.1.3 Pathogenesis

There are long-term neurological complications due to the alphavirus infection. Incubation period ranges from 1 to 10 days. In human host, transient viremia and dissemination might occur. Symptoms can be either fever, anxiety, headache (e.g., for EEE/WEE/VEE) or having fever, rash and arthralgia (e.g., Ross River, chikungunya and Sindbis viruses) [60].

3.8.1.4 Detection

Diagnosis is based on information from patients through clinical evidence or risk of exposure. Virus is detected by isolation from serum and detection by antibody-based method using IgG/IgM antibody against viral surface proteins.

3.8.1.5 Control and treatment

Seasonal human infection might happen, or it might be acquired in endemic areas. Vaccines are available. Controlling mosquitoes' populations or secondary host like horses when infected can lead to reduced spread to humans. Occasionally, the virus can escape its normal intermediate host and cause epizootic disease (VEE) or urban epidemics (chikungunya virus). Other treatment options include IFN, and antibodies help in clearance of the viruses.

3.8.2 Bunyaviridae

These are spherical, enveloped RNA viruses which could be either arboviruses (arthropod-borne) or roboviruses (rodent-borne). These have four genera: Bunyaviruses, Phlebovirus, Nairovirus, and Hantavirus. They further have 35 serogroups with nearly 300 types and subtypes.

3.8.2.1 Life cycle

They are 90–100 nm in diameter. The virion contains three segments of antisense ssRNA combined with nucleoprotein. After the viral genome enters the cytoplasm, they transcribe viral polymerase L protein and replicate. The genome also encodes four structural proteins, the glycoproteins Gc and Gn, and the nucleoprotein N. Few bunyaviruses have ambisense genome (those in which both nucleic acid strands encode for proteins) to encode the nonstructural proteins. They bud out of the cells into vesicles at or near the Golgi apparatus [61].

3.8.2.2 Pathogenesis

The virus causes disease in both human and domestic animals. Common symptoms include headache, nausea, body pain, conjunctival injection and generalized fragility followed by acute symptoms like hemorrhagic fever, renal failure, encephalitis and meningitis. Arthropods are intermediate carriers. Except for hantaviruses, biologic transmission occurs by a tick, mosquito, midge or sandfly vector. Humans are usually dead-end hosts for most of these viruses. Many of these viruses are known to cause specific disease like hemorrhage caused by Crimean-Congo hemorrhagic fever virus; encephalitis by La Crosse virus and related viruses, hemorrhagic hepatitis, encephalitis or blindness by Rift Valley fever virus infection [62].

3.8.2.3 Detection

Viral RNA can be detected in sera qRT-PCR. Further, specific IgMs/IgGs can be detected in sera of patients.

3.8.2.4 Control and treatment

Vaccination is available for most of these viruses (for humans to protect from Crimean-Congo and hantavirus hemorrhagic fever, and for sheep and cattle to protect against Rift Valley fever).

3.8.3 Flaviviruses

These viruses belong to family Flaviviridae and include the West Nile virus, dengue virus, tick-borne encephalitis virus, yellow fever virus and Zika virus, which cause a wide range of diseases that cause encephalitis.

3.8.3.1 Life cycle

They are ssRNA viruses with linear nonsegmented genomes with only one single ORF. The genome of these viruses is around 11,000 nucleotides, which encode around 10 genes. The virus capsid is icosahedral shaped, 40–50 nm in diameter and with a spherical-shaped envelope. They encode seven nonstructural proteins that include protease, helicase, replicase and virion maturation [63].

3.8.3.2 Pathogenesis

Many of these viruses induce the formation of liposome-mediated vesicle packets (VP), ER-derived-membranous structures containing viral dsRNA and viral NS proteins. So, these viruses utilize the host system to increase liposome synthesis, e.g., the cholesterol-synthesizing enzyme HMGCR is upregulated by the West Nile virus resulting in redistribution of cholesterol from VP disrupting the lipid raft domains. Most flavivirus has three structural proteins: the capsid proteins (C/V2), the matrix proteins (M/V1) and the envelope proteins and gps (E/V3) [64].

3.8.3.3 Detection

Viral RNA can be detected by qRT-PCR or nested PCR and by using specific antibody against the virus.

3.8.3.4 Control and treatment

Inactivated vaccines are available for both yellow fever virus and Japanese encephalitis. Since these viruses are mostly transmitted by arthropods like mainly mosquitoes and ticks, the best ways to prevent infection involve precautionary measures, such as using mosquito nets, bug-repellent and protective clothing or prevent infected body fluid/blood contact [65].

3.9 Retrovirus and human immunodeficiency virus

Retroviruses are ssRNA viruses and replicate their genome through a dsDNA intermediate.

3.9.1 Life cycle

Most retroviruses have 9–10 kb RNA genome. The reverse transcriptase protein encoded by the virus initiates creation of a dsDNA of the genome resulting in integration into the host cell chromosome. Retroviruses require three genes for replication: gag (structural proteins), pol (replication enzyme) and env (envelope gps that help in attachment) [66, 67].

3.9.2 Pathogenesis

Most retroviruses result in immortalization of the cells by expressing an oncogene, including Rous sarcoma virus and mouse mammary tumor virus. However, non-transforming viruses like lentiviruses, including HIV, do not transform cells. Instead, they replicate inside the cell and lyse out causing cell death in some cell types in which they replicate. Damaging the cells of the immune system results in losing the ability of an infected individual to mount an effective immune response, thus being susceptible to many other diseases [68].

3.9.3 Human immunodeficiency virus (HIV)

HIV is a lentivirus that causes acquired immunodeficiency syndrome (AIDS). Two well-characterized HIV types are HIV-1 and HIV-2. HIV-1 is the most virulent and accounts for majority of HIV infections. HIV-2 has lower infectivity.

3.9.3.1 Life cycle

The virus enters the cell by binding to the trimeric envelope complex (gp160 spike) with a CD4 and a chemokine receptor (CCR5 or CXCR4) on the cell surface. After the fusion of gp120 to CD4, the virus undergoes structural change, allowing gp120 to interact the target chemokine receptor. Thereafter, the HIV RNA and various enzymes, including reverse transcriptase, integrase, ribonuclease, and protease, are released into the cell. The RNA of these viruses has at least seven structural genes and nine enzymatic genes (gag, pol, env, tat, rev, nef, vif, vpr, vpu and tev). The RNA is reverse transcribed into dsDNA by a virally encoded reverse transcriptase. The resulting viral DNA goes to nucleus and integrates with cellular DNA by a virally encoded integrase enzyme. Once integrated, the virus enters the latent phase, multiplying slowly and avoiding recognition by the host immune system. In case the virus lyses out, its envelope contains both proteins from the host cell and HIV envelope protein [69].

3.9.3.2 Pathogenesis

HIV mostly infects the immune cells such as CD4+ T cells, macrophages and micro-glial cells. Initial infection results in flu-like illness within 2–4 weeks or people re-main asymptomatic for decade or longer. HIV infects the immune cells so that body cannot fight off infections and disease. Thereafter, compromised immune system results in AIDS, the last stage of HIV infection. They get a lot of opportunis-tic infections, including cryptococcal meningitis, toxoplasmosis, PCP (a type of pneumonia), esophageal candidiasis and certain cancers, including Kaposi's sar-coma (KS). Other symptoms during the last stage are chills, fever, sweats, swollen lymph glands, weakness and weight loss [70].

3.9.3.3 Detection

HIV-infected people can be monitored using two tests: (a) counting the CD4 cells. The immune cells are used by HIV to multiply, and they lyse out of the cells reduc-ing their number. When the cells reach fewer than 200 cells per mm^3, the immune system is severely compromised and does not function adequately; (b) determina-tion of viral load directly measures the amount of virus in the blood [71, 72].

3.9.3.4 Control and treatment

HIV can be controlled by proper medical care. Antiretroviral therapy (ART) can be used, which mainly includes nucleoside reverse transcriptase inhibitors like lami-vudine, emtricitabine, nevirapine, delavirdine, efavirenz, rilpivirine and integrase strand transfer inhibitors such as raltegravir and elvitegravir. HIV can transfer through blood-to-blood contact. The precautions include having protected sex, i.e., with a condom, not involved in drug abuse, i.e., not using needle sharing, mother to baby transfer can happen, so precautions should be taken [73].

3.10 Human herpesvirus (HHV)

Herpesviridae have dsDNA genome in an icosapentahedral capsid enclosed by a gp-bearing lipid bilayer envelope with length from 120 to 230 kbp having 31–75% G + C content and contain 60–120 genes. They have three groups: **α herpesviruses** include herpes simplex virus (HSV) types 1 and 2, and varicella-zoster virus; **β herpesviruses** include cytomegalovirus (CMV) and human herpesviruses (HHVs) 6 (HHV6) and 7 (HHV7); and **γ herpesviruses** which include Epstein–Barr virus (EBV) and HHV8.

3.10.1 Human herpesvirus 1 (HHV1)

HHV1 disseminates by skin-to-skin contact, typically causing cold sores around the mouth. Also known as HSV1, they can lead to infection in the genital area through oral–genital contact, causing genital herpes. It is likely to spread directly from an infected person.

3.10.2 Human herpesvirus 2 (HHV2)

It is a sexually transmitted infection. Also known as HSV2, it causes genital herpes. Sometimes they are known to cause cold sores in the facial area. Infection can also spread by skin-to-skin contact and the virus does not survive very long outside the body.

3.10.3 Human herpesvirus 3 (HHV3)

HHV3 causes chickenpox resulting in recurrent infection of the skin by varicella zoster virus. Shingles occur when the dormant varicella zoster virus after chickenpox becomes reactivated and infects skin cells and nerve cells.

3.10.4 Human herpesvirus 4 (HHV4, Epstein–Barr virus)

HHV4 is known to cause a contagious infection called infectious mononucleosis. This disease is also known as "mono" – the "kissing disease." It is transmitted from an infected person through body fluid contact like coughing, sneezing or sharing common utensils.

3.10.5 Human herpesvirus 5 (HHV5)

Also known as CMV is known to cause mononucleosis. It can be transmitted sexually, even to newborns through breast-feeding, blood transfusions and organ transplants, and can also cause hepatitis. CMV can easily infect a person with compromised immune system like AIDS. It causes diarrhea, vision problems resulting in blindness, stomach and intestine infections and even death.

3.10.6 Human herpesvirus 6

HHV6 causes roseola in small children with high fever and skin rash, and convulsions associated with fever.

3.10.7 Human herpesvirus 7

HHV7 is very common and can also cause roseola, but other clinical effects are unclear.

3.10.8 Human herpesvirus 8

This is also known as KS virus, known to cause tumors in patients with AIDS.

3.10.9 Multiplication

Gps present in viral envelope bind to cell membrane receptors and internalize the virion. The viral DNA is released in the cell cytoplasm which then migrates to the nucleus. Viral replication takes place inside the nucleus. Herpesviruses hijack the host's transcription machinery and repair enzymes to support its replication. Finally, the virion buds out through the nuclear membrane or endoplasmic reticulum leading to the death of host cells. However, in some host cells, the virus can persist in the cell by integrating with host DNA [74].

3.10.10 Detection

Most of the herpes infection can be detected based on clinical symptoms. Others rely on detection of viral genes or gene products, and isolation of the virus through culturing and detecting viral DNA/RNA using qRT-PCR.

3.10.11 Control and treatment

Live-attenuated vaccine against α-herpesvirus infection like chickenpox, varicella zoster, HSV2 and CMV are available. Antiviral drugs such as acyclovir, cidofovir, famciclovir, fomivirsen, foscarnet, ganciclovir, idoxuridine, penciclovir, trifluridine, valacyclovir, valganciclovir and vidarabine are also used. Immunoglobulin or hyperimmunoglobulin from sera are used for passive immunization in acute infections.

3.11 Human cancer virus

Viruses that cause cancer belong to a wide range of virus families, like the RNA virus, including Retroviridae and Flaviviridae, and the DNA virus including Hepadnaviridae, Herpesviridae and Papillomaviridae. These viruses can integrate with the host DNA, causing long-term persistent infections. Examples include (i) HTLV-1 (known to cause adult T-cell leukemia); (ii) human papillomavirus (known to cause cervical cancer, head and neck cancers and other anogenital cancers); (iii) HHV8 (known to cause KS, primary effusion lymphoma and Castleman's disease); (iv) EBV (known to cause Burkitt's lymphoma, nasopharyngeal carcinoma and Hodgkin's disease); and (v) HBV and HCV (known to cause hepatocellular carcinoma) [75].

3.11.1 Human T-cell leukemia virus (HTLV-1)

HTLVs were the first human retrovirus discovered and known to target T-lymphocytes. It is grouped in the family Retroviridae and genus Deltaretrovirus. It causes adult T-cell leukemia and HTLV-1–associated myelopathy.

3.11.1.1 Life cycle

As the viral RNA along with other proteins enters the host cytoplasm, its reverse transcriptase enzyme generates proviral DNA, which is then integrated into the host genome by viral integrase. Afterward, it remains latent and spreads in the dividing cells. Continuous expression of viral oncogenes replication machinery in T-lymphocyte results in its increased proliferation. HTLV-1 targets CD4 lymphocytes, while HTLV-2 affects CD8 lymphocytes [76].

3.11.1.2 Pathogenesis

HTLV-1 can be transmitted through blood or body fluids like mother to child through breastfeeding, sexual intercourse and sharing of needles and syringes. Similar to HIV, they also have a gag–pol–env motif with flanking long terminal repeat sequences, but they also have a fourth sequence named Px, which encodes for regulatory proteins Tax, Rex, p12, p13 and p30 and which helps in both infection and viral replication. Chronic infection results in general weakness and lack of energy; skin, bone and multiple visceral lesions or pulmonary infiltration; and hepatosplenomegaly and hypercalcemia [77].

3.11.1.3 Detection

EIA or particle agglutination assays are commonly used to detect antibodies against HTLV-1 in the peripheral blood of a patient with symptoms. General effects include skin lesions, high leukocyte counts and CD25+ cells.

3.11.1.4 Control and treatment

These patients have poor prognosis because of the multidrug resistance of malignant cells, a large tumor burden with multiorgan failure. T-cell immunodeficiency results in increased rate of infection. HTLV-1 infection may be prevented avoiding blood-to-blood or body fluid contact with the infected individual. The major prognostic factors are advanced age, high calcium or lactic dehydrogenase levels and active infections or lesions [78].

3.11.2 Human papillomavirus

HPVs contain more than 200 related viruses that can be divided into two groups: low-risk HPVs and high-risk HPVs. Low-risk HPVs cause skin warts. High-risk HPVs, like HPV types 16 and 18, are known to cause most cases of cervical cancer (>90%), 30% of head and neck cancer, 65% of vaginal cancers, 50% of vulvar cancers and 35% of penile cancers [79].

3.11.2.1 Life cycle

The virions initially attach to the basal membrane of the epithelium with the heparan sulfate proteoglycan, while they multiply and shed out of the terminally differentiated cells in the cornified layer at the surface of the epithelium. After infection, the virus maintains a low-copy number. There are two major promoters: the early promoter and the late promoter. The replication of the virus is dependent on differentiation of the epithelium [80].

3.11.2.2 Pathogenesis

HPV infection generally self-heals. However, when the infection persists, it results in genital warts or certain type of cancers. The two well-studied oncogenic proteins are E6 and E7, which play an important role in virulence of the virus. High-risk HPV E6 proteins form a complex with E6-associated protein that targets the p53 tumor suppressor protein for degradation. Moreover, there are various p53-independent activities of E6 like telomerase activation, association with proteins having PDZ-binding protein and other cellular proteins. High-risk HPV E7 proteins degrade the retinoblastoma tumor suppressor protein causing activation of E2F transcription factors resulting in dysregulated cell cycle [81].

3.11.2.3 Detection

Genital warts are small, flat or cauliflower-shaped, small bump, cluster of bumps or stem-like protrusions. Other common warts, plantar and flat warts have distinct morphology. Most common tests for detecting HPV cancer can be as simple as the use of acetic acid (vinegar), or by Pap smear and a DNA test. In the first two methods, any cellular abnormalities that may lead to cancer can be detected from the cells at the surface of the cervix or the vagina in women. However, PCR, viral RNA detection and p16 immunostaining are confirmatory tests for HPV infection [82].

3.11.2.4 Control and treatment

Warts can clear by themselves or by using over-the-counter salicylic acid, podophyllin, imiquimod (Aldara and Zyclara), etc. Various other methods of treatment include cryosurgery (freezing that destroys tissue), loop electrosurgical excision procedure, surgical conization from the cervix and cervical canal, and laser vaporization conization. Infection can also be prevented by stopping blood and fluid contact with diseased individual, abstinence, monogamous sexual relationships and the use of HPV vaccines. Three well-known HPV vaccines available are quadrivalent vaccine Gardasil, bivalent vaccine Cervarix and nonavalent vaccine Gardasil 9 [83].

3.11.3 Epstein–Barr virus (EBV)

EBV is a B-lymphotropic herpesvirus discovered in 1964. Two major types are type A and type B. They differ only in the genes that encode some of the nuclear proteins like EBV nuclear Ag, EBNA-2, EBNA-3A, EBNA-3B and EBNA-3C. It has a linear, 172 kb dsDNA genome. EBV is known to cause certain lymphomas and nasopharyngeal cancer [84].

3.11.3.1 Life cycle

EBV's life cycle consists of two phases: it replicates in host cells and causes cell lysis, the so-called lytic phase, or it can integrate into the host cell DNA causing the latent phases. The latent mode of replication is most common, and horizontal transfer occurs by host cell replication. Two genes that mediate the switch between latency and the lytic cycle of EBV, BZLF1 and BRLF1 encode transcriptional activator proteins [85].

3.11.3.2 Pathogenesis

EBV infects with B cells in humans by interacting with viral envelope gp 350 with the cell surface receptor CR2. The virus is then endocytosed by fusion with the host cell membrane. This mechanism involves three viral gps: gp85, gp25 and gp42. EBV undergoes a nonproductive lysogenic cycle with the expression of restricted number of viral genes. Symptoms are usually mild and self-limited, but the virus can persist. In some cases, acute mononucleosis causes fatigue [86].

3.11.3.3 Detection

Antibodies to EBV are highly cross-reactive to other EBV homologues, so antibody-based detection is not effective. EBV DNA is detected in the nuclei of blood lymphocytes of the infected person. In infected people, the number of normal B cells in the blood is increased and the cells look unusual or "atypical" under the microscope. Few might develop anemia due to the hemolysis of the infected RBCs and might also have a low platelet count [87].

3.11.3.4 Control and treatment

Mononucleosis is also known as "kissing disease." Mononucleosis cannot be treated. However, corticosteroids can be used to reduce symptoms like swelling in the throat or an enlarged spleen. Antiviral medications including acyclovir, ganciclovir and foscarnet are used to treat certain symptoms. Exchange of body fluids should be prevented like kissing or sharing food.

3.11.4 Kaposi's sarcoma-associated herpesvirus (KSHV)

Also known as HHV8, it is DNA gamma-herpesvirus known to cause KS, primary effusion B-cell lymphoma and other lymphoproliferative disorder like Castleman's disease.

3.11.4.1 Life cycle

Both lytic and lysogenic life cycle are observed in these viruses. After initial infection, the virus goes to a latent phase with episomal DNA encoding early transcripts including v-FLIP, v-cyclin, kaposins A, B, C and nuclear Ag (LANA-1). Activation of late transcripts results in the activation of the lytic cycle [88].

3.11.4.2 Pathogenesis

It can be spread through body–fluid contactor by ways like organ transplantation and breastfeeding. Mostly early infections are nonsymptomatic. KSHV (Kaposi's sarcoma-associated herpesvirus) infects monocytes, dendritic cells, B lymphocytes, oral epithelial cells and endothelial cells. Types of KS include:

(a) **Classic KS** infects older people with symptoms like slow growing lesions usually on the skin of ankles and soles of the feet.

(b) **Epidemic KS** is also known as AIDS-associated KS, which is more aggressive with several skin lesions.

(c) **Endemic KS** is called African KS, which can affect both children and adults.

(d) **Iatrogenic KS** is an infection that occurs due to treatment of immunosuppressive drugs and during transplantation [89].

3.11.4.3 Detection

Any kind of tumors in person with AIDS or those under immune-suppressed drugs during transplantation are due to KS. Tumors can also be examined by computerized tomography scans, bronchoscopy lung examination and gastrointestinal endoscopy examination.

3.11.4.4 Control and treatment

Classic KS can be treated with imiquimod cream as local treatment for superficial tumors. Highly active ART in AIDS patients helps manage KS infections. Tumors can be removed by surgical excision, cryotherapy and electrocoagulation [90].

3.12 Rabies virus

Rabies virus is a rod-shaped negative-stranded RNA virus of the family Rhabdoviridae, order Mononega virales and three genera, Lyssa virus, Ephemero virus and Vesiculo virus.

3.12.1 Life cycle

Rhabdoviruses are around 180 nm long and 75 nm wide with knob-like spikes in the outer membrane composed of gp G, which allow both receptor binding and membrane fusion. The virus genome ranges from 11,615 to 11,966 nucleotide and encodes five proteins that code for nucleoprotein (N), phosphoprotein (P), matrix protein (M), glycoprotein (G) and RNA polymerase (L) [91, 92].

3.12.2 Pathogenesis

In case the virus enters through the skin or mucous membrane through a wound site, the virus can replicate in the striated muscles. From muscle cells, it can migrate to the nervous system via unmyelinated sensory nerve endings. The rabies virus can multiply in neuronal cells, so it often targets the nervous tissue. Five steps of infection are incubation, prodrome, acute neurologic period, coma and death. The incubation period is highly variable, usually 1–3 months till it reaches the brain and spreads throughout the body via efferent neural pathways. The symptoms range from fever, and fatigue, or respiratory disorders like sore throat, cough and dyspnea, and gastrointestinal disorders like abdominal pain and diarrhea or affects central nervous systems causing vertigo, anxiety, apprehension, irritability and nervousness in acute cases [93].

3.12.3 Detection

Quantitative real-time PCR can be used to detect viral DNA in saliva. Rabies virus antibodies may be tested in serum and spinal fluid. Rabies Ag can also be tested in the cutaneous nerves at the base of hair follicles during skin biopsy using the direct fluorescent antibody test test.

3.12.4 Control and treatment

Transmission can be from the infected saliva of a to an uninfected animal through direct contact such as a scratch or other break in the skin or bite. Live human diploid cell vaccine (HDCV) is available. Passive immunization is given to patients who have not received vaccine with rabies immune globulin, either human or equine type, followed by the active, HDCV shot. An adsorbed rabies vaccine (RVA) is also available. Pet dogs and cats should be vaccinated.

3.13 Coronavirus

It is derived from the Latin corona, meaning halo, which refers to the appearance of virions. They belong to the order Nidovirales, which have three families: Coronaviridae, Arteriviridae and Roniviridae. In high-resolution EM, the virus looks like a fringe of large, spherical surface projections similar to a royal crown. They are further of four groups: the alpha, beta, gamma and delta coronaviruses. The six well-known coronaviruses that infect humans are alpha coronaviruses (229E and NL63) and beta coronaviruses (OC43, HKU1, SARS (severe acute respiratory syndrome)-CoV and MERS-CoV) [94]. SARS-CoV (COVID-19) is responsible for the recent coronavirus breakout. It belongs to the genus Betacoronavirus.

3.13.1 Life cycle

Viral outer membrane has spike (S) protein and its receptor on the host surface. After fusion with the cell, the virus genome enters the cell. The virus has a ~30 kb positive-sense RNA genome, which codes for four proteins: the spike (S), membrane (M), envelope (E) and the nucleocapsid (N) protein found in the core. The replicase gene is encoded by the two-thirds of the viral genome. The ~150 kDa S protein has an N-terminal signal sequence targeted for endoplasmic reticulum and N-linked glycosylated like HA esterase in some betacoronaviruses. New virions form by budding from host cell membranes. The E protein is thought to be critical for coronavirus infectivity [95].

3.13.2 Pathogenesis

Transmission is via airborne droplets to the nasal mucosa. Other method of transmission is spray-based aerosolization in a confined space or contamination of common surfaces. SARS-CoV belongs to 2b β-corona virus known to cause SARS outbreak in 2002–2003. It targets epithelial cells within the lungs. They enter the host cells through interacting with ACE2 receptors present on type 2 pneumocytes, intestinal epithelial cells and nasal goblet secretory cells. Other coronaviruses like α-corona viruses (HCoV-229E and HCoV-NL63) and β-coronaviruses (HCoV-OC43 and HCoV-HKU1) are known to cause mild, self-limiting respiratory infections in humans. MERS-CoV is a zoonotic virus, which can spread by contact with infected camels [96].

3.13.3 Detection

The virus is difficult to isolate; however, antibodies are present in the serum. Nucleic acid hybridization tests and viral DNA PCR from saliva and nasal fluids are available now. Pneumonia is a common symptom; however, gastrointestinal symptoms, including diarrhea, fever, cough and shortness of breath, may appear in acute cases.

3.13.4 Control and treatment

Generally, symptoms subside after 1–2 weeks. Immunocompromised patients may require ventilation support and symptomatic medications. Vaccines against coronavirus are currently available. IFN can protect against infection. Current CDC guidelines recommend physical distancing by 6 feet away and mask wearing to reduce the risk of exposure. Surface disinfection methods like using 60–70% ethanol or 0.5% hydrogen peroxide also help in reducing exposure.

3.14 Rubella virus

Rubella virus causes 3-day measles or also known as German measles. It is a communicable viral infection. The virion is 40–80 nm spherical, with positive-sense, ssRNA. It has spike-like, HA-containing surface projections on its surface.

3.14.1 Life cycle

The virus replicates in the cytoplasm. The genomic RNA is capped and polyadenylated and produces nonstructural proteins. The virus forms a polyprotein that is proteolytically cleaved into an RNA polymerase that helps in the replication of the virus. The polyprotein codes for three major structural polypeptides: two gps that are present on the membrane, E1 encoding for viral hemagglutination and neutralization; and E2, a glycosylated protein along with a single nonglycosylated RNA-associated capsid protein, C, within the virion [97].

3.14.2 Pathogenesis

Rubella virus attaches and multiplies in respiratory epithelium in trachea and bronchi. This often leads to postauricular lymphadenopathy, and occipital and posterior cervical lymph nodes. Viremia happens in 5–7 days with symptoms appearing within 1–2 weeks. Postnatal rubella is often mild, resulting in rash, lymphadenopathy and low-grade fever. Another form, congenital rubella, results in swollen glands or lymph nodes with high fever. Infected persons may also develop arthritis and painful joints [98, 99].

3.14.3 Detection

Clinical symptoms is a typical rash and lymphadenopathy with Forchheimer's sign in few cases. The latter is characterized by small, red papules on the area of the soft palate. Antibody-based detection with serum of infected person confirms the presence of virus in >90% of cases within 5 days after the onset of fever and rash. Rubella RNA can also be detected by quantitative real-time PCR and endpoint PCR.

3.14.4 Control and treatment

Live-attenuated rubella vaccine is available as measles vaccine (measles and rubella) or measles and mumps vaccines (measles, mumps and rubella). The immunity is generally lifelong. Immunoglobulin treatment is often used for pregnant women.

3.15 Prion

Prions are known to cause neurodegenerative diseases by converting normal prion protein (PrPC, prion-related protein) into an abnormal form (PrPSc, in which Sc stands for scrapie, the prion disease of sheep and goats). The normal prion has α-helical structure; however, the abnormal protein has β-pleated sheet structure. German neuropathologists Alfons Maria Jakob and Hans Gerhard Creutzfeldt described the first cases of prion infection between 1921 and 1923 [100].

3.15.1 Life cycle

Prion-related protein (PrP) is a 253 amino acid protein that attaches to the outer surface of cell membranes. Both the normal and abnormal proteins are encoded by

PRNP gene of human chromosome 20 (20p13). Although they have similar primary amino acid sequence, however, they differ only in secondary and tertiary structures. The normal protein is predominantly rich in alpha helical contents, while the misfolded protein is predominantly rich in beta sheet contents [101].

3.15.2 Types

The misfolding can happen sporadically, or it can be formed by hereditary or acquired mechanisms. Examples of sporadic cases include Creutzfeldt–Jakob disease (CJD), fatal insomnia and variably protease-sensitive prionopathy. Besides CJD could be familial or genetic, other examples include fatal familial insomnia (FFI) and Gerstmann–Sträussler–Scheinker (GSS) syndrome. Kuru and iatrogenic CJD (iCJD) are examples of acquired human prion diseases.

3.15.2.1 Sporadic Creutzfeldt–Jakob disease (sCJD)

Sporadic CJD (sCJD) is most common with around 85% cases. It is a neurodegenerative disorder, rapidly progressive and mostly fatal within 1 year.

A polymorphism in the prion gene, PRNP, is observed at codon 129 in case of sCJD, which can be either methionine (M) or valine (V). Clinical diagnosis involves neurological symptoms and can be detected using antineuronal autoantibody in cerebrospinal fluid, computed tomography, electroencephalography (EEG) and magnetic resonance imaging (MRI). In case of sCJD, two most common diagnostic markers are the presence of 14-3-3 protein in the cerebrospinal fluid and typical electroencephalogram (EEG) pattern.

The classic symptoms include dementia, which is rapidly progressive with abnormalities in behavior, ataxia and extra pyramidal features. Notably, sCJD symptoms are like Alzheimer's disease, dementia or paraneoplastic syndrome. So histopathologic changes in biopsy or autopsy of brain tissue showing nerve cell loss, gliosis, vacuolation (formerly called spongiform change) and PrPSc deposition indicate a positive sCJD case [102, 103].

3.15.2.2 Variant Creutzfeldt–Jakob disease

Variant CJD is known to cause bovine spongiform encephalopathy ("mad cow" disease), a prion disease in cow. Human epidemic might be due to feeding animal products specifically sheep or cow brain contaminated with the prion disease scrapie. The age of onset is 29 years with mean duration of disease spanning 18 months.

Symptoms usually begin with a psychiatric complication. This is followed by neurological disorder, cognitive dysfunction, dysesthesia, cerebellar dysfunction and involuntary movements. The presence of kuru-type amyloid plaques surrounded by spongiform lesions called "the florid plaques" is often seen in histopathologic findings. The genetic change involves homozygous substitution of methionine at PRNP codon 129. The age of onset is 27 years with mean duration of disease spanning 14.5 months. Late-onset signs include irregular urination, progressive immobility and akinetic mutism leading to death in some cases [104].

3.15.2.3 Iatrogenic Jakob–Creutzfeldt disease

iCJD was first observed in a person who received cornea from a patient having CJD in 1974. Iatrogenic transmission happens by direct contact with patient proteins, i.e., by using intracerebral EEG needles, cadaveric grafts and injections of contaminated cadaveric pituitary-derived human growth hormone and gonadotrophin hormone. The clinicopathologic features are like kuru. An additional risk factor is homozygosity 129MM with incubation from 4.5 to >25 years [105].

3.15.2.4 Kuru

Kuru can be identified through histopathologic identification of peculiar plaques in brain tissue looking like a spherical body with a rim of spreading filaments. This is due to a heterozygous mutation in the human prion protein gene (PRNP) especially at codons 127 and 129. Onset starts with headache and pain usually in the joints of legs. This is followed by cerebellar ataxia, tremors and choreiform. Complication increases in cold. There are three clinical stages in kuru based on clinicopathologic characteristics observed in patient: they can still walk (ambulant), they can only sit up (sedentary) and they are unable to sit up independently (terminal) [106].

3.15.2.5 Fatal familial insomnia (FFI)

FFI has *cis*-methionine mutation at codon 129. FFI occurs equally in men and women. Early symptoms involve patients with severe progressive insomnia over several months, followed by tachycardia, hyperhidrosis and hyperpyrexia. Late symptoms involve manifestations of motor and cognitive disorder including myoclonus, ataxia, dysarthria, dysphagia and pyramidal signs, and autonomic hyperactivation. The age of onset is 27 years with mean duration of disease spanning 18 months [107].

3.15.2.6 Gerstmann–Sträussler–Scheinker (GSS) syndrome

GSS is due to autosomal-dominant inheritance of PRNP mutations with an annual prevalence of one to two cases per million of populations. Besides M129V mutation, other mutations that are cofound in GSS are P102L, P105L, A117V, Y145X, Q160X, F198S, Q217R, Y218N, Y226X and Q227X. They are slow in progression and have an early onset between 30 and 60 years. The early symptoms are brain disorders, cerebellar ataxia, abnormalities in posture and walking, dementia, dysarthria, ocular dysmetria, signs of parkinsonian and disorder of the lower extremities. Diagnosis can be done by MRI and improved CSF tests like positron emission tomography. Increased PrP deposits are observed in the cerebellum with a strip-like pattern and in the subiculum entorhinal region [108, 109].

3.16 Conclusion

Viruses consist of nucleic acids which could be DNA or RNA. They are surrounded by a protein coat known as a capsid made of basic unit called capsomere. The ICTV system and Baltimore classification system are two major systems for viral classification. The International Union of Microbiological Societies developed the ICTV system to refine and maintain a universal virus taxonomy. Viruses are obligate intracellular parasites and face continuous environmental insult from cell to cell and therefore keep on changing because of genetic selection. Viruses show enormous variations in their epidemiology, symptoms and pathogenesis. Many such infections are stopped by our innate immunity. The first line of defense is induction of type I and III IFNs involved in innate immune responses. Further, every year many new viral strains emerge as they undergo many genetic changes through mutation and recombination. However, viral infection can be reduced by limiting exposure, by eliminating nonhuman reservoirs, by eliminating the vector and by improving sanitation. The presence or absence of the active viral infection in a clinical setting is by identifying at its DNA/RNA level, i.e., viral load or using microscopy to visualize virion particles or by using its corresponding antibodies, to prove active infection or past exposure. Vaccines provide long-term immunity to many viral infections.

References

[1]   Davison AJ, Journal of general virology – introduction to 'ICTV Virus Taxonomy Profiles', J Gen Virol, 2017, 98, 1.
[2]   Baltimore D, Expression of animal virus genomes, Bacteriol Rev, 1971, 35, 235–41.

[3] Sanjuán R, Domingo-Calap P, Mechanisms of viral mutation, Cell Mol Life Sci, 2016, 73, 4433–48.

[4] Sanjuan R, Domingo-Calap P, Mechanisms of viral mutation, Cell Mol Life Sci: CMLS, 2016, 73, 4433–48.

[5] Steinhauer DA, Domingo E, Holland JJ, Lack of evidence for proofreading mechanisms associated with an RNA virus polymerase, Gene, 1992, 122, 281–88.

[6] Harris RS, Dudley JP, APOBECs and virus restriction, Virology, 2015, 479–480, 131–45.

[7] Lynch M, Evolution of the mutation rate, Trends Genet, 2010, 26, 345–52.

[8] Johansson C, Schwartz S, Regulation of human papillomavirus gene expression by splicing and polyadenylation, Nat Rev Microbiol, 2013, 11, 239–51.

[9] Halliburton IW, Randall RE, Killington RA, Watson DH, Some properties of recombinants between type 1 and type 2 herpes simplex viruses, J Gen Virol, 1977, 36, 471–84.

[10] Boni MF, de Jong MD, van Doorn HR, Holmes EC, Guidelines for identifying homologous recombination events in influenza A virus, PloS One, 2010, 5, e10434.

[11] Chetverin AB, The puzzle of RNA recombination, FEBS Lett, 1999, 460, 1–5.

[12] McVean G, Awadalla P, Fearnhead P, A coalescent-based method for detecting and estimating recombination from gene sequences, Genetics, 2002, 160, 1231–41.

[13] Taucher C, Berger A, Mandl CW, A trans-complementing recombination trap demonstrates a low propensity of flaviviruses for intermolecular recombination, J Virol, 2010, 84, 599–611.

[14] Rabadan R, Levine AJ, Krasnitz M, Non-random reassortment in human influenza A viruses, Influenza Other Respir Viruses, 2008, 2, 9–22.

[15] Bernstein C, Deoxyribonucleic acid repair in bacteriophage, Microbiol Rev, 1981, 45, 72–98.

[16] Haller O, Weber F, Pathogenic viruses: smart manipulators of the interferon system, Curr Top Microbiol Immunol, 2007, 316, 315–34.

[17] Wilkins C, Gale M Jr, Recognition of viruses by cytoplasmic sensors, Curr Opin Immunol, 2010, 22, 41–47.

[18] Balachandran S, Roberts PC, Brown LE, et al., Essential role for the dsRNA-dependent protein kinase PKR in innate immunity to viral infection, Immunity, 2000, 13, 129–41.

[19] Lokshin A, Mayotte JE, Levitt ML, Mechanism of interferon beta-induced squamous differentiation and programmed cell death in human non-small-cell lung cancer cell lines, J Natl Cancer Inst, 1995, 87, 206–12.

[20] Tsushima H, Imaizumi Y, Imanishi D, Fuchigami K, Tomonaga M, Fas antigen (CD95) in pure erythroid cell line AS-E2 is induced by interferon-gamma and tumor necrosis factor-alpha and potentiates apoptotic death, Exp Hematol, 1999, 27, 433–40.

[21] Jo M, Kim TH, Seol DW, et al., Apoptosis induced in normal human hepatocytes by tumor necrosis factor-related apoptosis-inducing ligand, Nat Med, 2000, 6, 564–67.

[22] Kotenko SV, IFN-lambdas, Curr Opin Immunol, 2011, 23, 583–90.

[23] Yoneyama M, Fujita T, RNA recognition and signal transduction by RIG-I-like receptors, Immunol Rev, 2009, 227, 54–65.

[24] Golden-Mason L, Rosen HR, Natural killer cells: multifaceted players with key roles in hepatitis C immunity, Immunol Rev, 2013, 255, 68–81.

[25] Ljunggren HG, Karre K, In search of the 'missing self': MHC molecules and NK cell recognition, Immunol Today, 1990, 11, 237–44.

[26] Zimmermann P, Curtis N, Antimicrobial effects of antipyretics, Antimicrob Agents Chemother, 2017, 61.

[27] Graham BS, Sullivan NJ, Emerging viral diseases from a vaccinology perspective: preparing for the next pandemic, Nat Immunol, 2017.

[28] Huang HS, Tsai CL, Chang J, Hsu TC, Lin S, Lee CC, Multiplex PCR system for the rapid diagnosis of respiratory virus infection: a systematic review and meta-analysis, Clin

Microbiol Infect, the official publication of the European Society of Clinical Microbiology and Infectious Diseases, 2017.

[29]   Zhao M, Wang L, Li S, Influenza A virus-host protein interactions control viral pathogenesis, Int J Mol Sci, 2017, 18.

[30]   Cline TD, Beck D, Bianchini E, Influenza virus replication in macrophages: balancing protection and pathogenesis, J Gen Virol, 2017, 98, 2401–12.

[31]   Villalon-Letelier F, Brooks AG, Saunders PM, Londrigan SL, Reading PC, Host cell restriction factors that limit Influenza A infection, Viruses, 2017, 9.

[32]   Agrawal P, Nawadkar R, Ojha H, Kumar J, Sahu A, Complement evasion strategies of viruses: an overview, Front Microbiol, 2017, 8, 1117.

[33]   de Boer PT, van Maanen BM, Damm O, et al., A systematic review of the health economic consequences of quadrivalent influenza vaccination, Expert Rev Pharmacoecon Outcomes Res, 2017, 17, 249–65.

[34]   La Rosa G, Fratini M, Della Libera S, Iaconelli M, Muscillo M, Viral infections acquired indoors through airborne, droplet or contact transmission, Ann Ist Super Sanita, 2013, 49, 124–32.

[35]   Crawford SE, Ramani S, Tate JE, et al., Rotavirus infection, Nat Rev Dis Primers, 2017, 3, 17083.

[36]   Royall E, Locker N, Translational control during calicivirus infection, Viruses, 2016, 8, 104.

[37]   Blacklow NR, Norwalk virus and other caliciviruses, 4th edn, Samuel Baron, eds., Medical Microbiology, University of Texas Medical Branch at Galveston (TX), 1996.

[38]   Lion T, Adenovirus infections in immunocompetent and immunocompromised patients, Clin Microbiol Rev, 2014, 27, 441–62.

[39]   Chen RF, Lee CY, Adenoviruses types, cell receptors and local innate cytokines in adenovirus infection, Int Rev Immunol, 2014, 33, 45–53.

[40]   Bosch A, Pinto RM, Guix S, Human astroviruses, Clin Microbiol Rev, 2014, 27, 1048–74.

[41]   Vaughan G, Goncalves Rossi LM, Forbi JC, et al., Hepatitis A virus: host interactions, molecular epidemiology and evolution, Infect Genet Evol, 2014, 21, 227–43.

[42]   Debing Y, Neyts J, Thibaut HJ, Molecular biology and inhibitors of hepatitis A virus, Med Res Rev, 2014, 34, 895–917.

[43]   Sunbul M, Hepatitis B virus genotypes: global distribution and clinical importance, World J Gastroenterol, 2014, 20, 5427–34.

[44]   Xu C, Zhou W, Wang Y, Qiao L, Hepatitis B virus-induced hepatocellular carcinoma, Cancer Lett, 2014, 345, 216–22.

[45]   Liu YP, Yao CY, Rapid and quantitative detection of hepatitis B virus, World J Gastroenterol, 2015, 21, 11954–63.

[46]   Manzoor S, Saalim M, Imran M, Resham S, Ashraf J, Hepatitis B virus therapy: what's the future holding for us? World J Gastroenterol, 2015, 21, 12558–75.

[47]   Ansaldi F, Orsi A, Sticchi L, Bruzzone B, Icardi G, Hepatitis C virus in the new era: perspectives in epidemiology, prevention, diagnostics and predictors of response to therapy, World J Gastroenterol, 2014, 20, 9633–52.

[48]   Dubuisson J, Cosset FL, Virology and cell biology of the hepatitis C virus life cycle: an update, J Hepatol, 2014, 61, S3–S13.

[49]   Mukherjee R, Burns A, Rodden D, et al., Diagnosis and management of hepatitis C virus infection, J Lab Autom, 2015, 20, 519–38.

[50]   Rizzetto M, Hepatitis D virus: introduction and epidemiology, Cold Spring Harb Perspect Med, 2015, 5, a021576.

[51]   Abbas Z, Afzal R, Life cycle and pathogenesis of hepatitis D virus: a review, World J Hepatol, 2013, 5, 666–75.

[52]   Alfaiate D, Deny P, Durantel D, Hepatitis delta virus: from biological and medical aspects to current and investigational therapeutic options, Antiviral Res, 2015, 122, 112–29.

[53]   Abbas Z, Abbas M, Management of hepatitis delta: need for novel therapeutic options, World J Gastroenterol, 2015, 21, 9461–65.

[54]   Kamar N, Dalton HR, Abravanel F, Izopet J, Hepatitis E virus infection, Clin Microbiol Rev, 2014, 27, 116–38.

[55]   Kumar S, Subhadra S, Singh B, Panda BK, Hepatitis E virus: the current scenario, Int J Infect Dis, 2013, 17, e228–33.

[56]   Cook N, van der Poel WH, Survival and elimination of hepatitis E virus: a review, Food Environ Virol, 2015, 7, 189–94.

[57]   Stapleton JT, Foung S, Muerhoff AS, Bukh J, Simmonds P, The GB viruses: a review and proposed classification of GBV-A, GBV-C (HGV), and GBV-D in genus Pegivirus within the family Flaviviridae, J Gen Virol, 2011, 92, 233–46.

[58]   Liang G, Li X, Gao X, et al., Arboviruses and their related infections in China: a comprehensive field and laboratory investigation over the last 3 decades, Rev Med Virol, 2017.

[59]   Rupp JC, Sokoloski KJ, Gebhart NN, Hardy RW, Alphavirus RNA synthesis and non-structural protein functions, J Gen Virol, 2015, 96, 2483–500.

[60]   Fros JJ, Pijlman GP, Alphavirus infection: host cell shut-off and inhibition of antiviral responses, Viruses, 2016, 8.

[61]   Briese T, Calisher CH, Higgs S, Viruses of the family Bunyaviridae: are all available isolates reassortants? Virology, 2013, 446, 207–16.

[62]   Beaty BJ, Calisher CH, Bunyaviridae – natural history, Curr Top Microbiol Immunol, 1991, 169, 27–78.

[63]   Ng WC, Soto-Acosta R, Bradrick SS, Garcia-Blanco MA, Ooi EE, The 5′ and 3′ untranslated regions of the flaviviral genome, Viruses, 2017, 9.

[64]   Pierson TC, Kielian M, Flaviviruses: braking the entering, Curr Opin Virol, 2013, 3, 3–12.

[65]   Heinz FX, Stiasny K, Flaviviruses and flavivirus vaccines, Vaccine, 2012, 30, 4301–06.

[66]   Mattei S, Schur FK, Briggs JA, Retrovirus maturation – an extraordinary structural transformation, Curr Opin Virol, 2016, 18, 27–35.

[67]   Peeters M, D'Arc M, Delaporte E, Origin and diversity of human retroviruses, AIDS Rev, 2014, 16, 23–34.

[68]   Lindemann D, Steffen I, Pohlmann S, Cellular entry of retroviruses, Adv Exp Med Biol, 2013, 790, 128–49.

[69]   Goodsell DS, Illustrations of the HIV life cycle, Curr Top Microbiol Immunol, 2015, 389, 243–52.

[70]   Kok YL, Ciuffi A, Metzner KJ, Unravelling HIV-1 latency, one cell at a time, Trends Microbiol, 2017, 25, 932–41.

[71]   Escolano A, Dosenovic P, Nussenzweig MC, Progress toward active or passive HIV-1 vaccination, J Exp Med, 2017, 214, 3–16.

[72]   Huynh K, Gossman WG, HIV, Prevention. Treasure Island (FL), StatPearls, 2017.

[73]   Kukhanova MK, Korovina AN, Kochetkov SN, Human herpes simplex virus: life cycle and development of inhibitors, Biochemistry (Mosc), 2014, 79, 1635–52.

[74]   Eisenberg RJ, Atanasiu D, Cairns TM, Gallagher JR, Krummenacher C, Cohen GH, Herpes virus fusion and entry: a story with many characters, Viruses, 2012, 4, 800–32.

[75]   Mui UN, Haley CT, Tyring SK, Viral oncology: molecular biology and pathogenesis, J Clin Med, 2017, 6.

[76]   Carpentier A, Barez PY, Hamaidia M, et al., Modes of human T cell leukemia virus type 1 transmission, replication and persistence, Viruses, 2015, 7, 3603–24.

[77] Satou Y, Matsuoka M, Virological and immunological mechanisms in the pathogenesis of human T-cell leukemia virus type 1, Rev Med Virol, 2013, 23, 269–80.

[78] Matsuoka M, Jeang KT, Human T-cell leukemia virus type 1 (HTLV-1) and leukemic transformation: viral infectivity, Tax, HBZ and therapy, Oncogene, 2011, 30, 1379–89.

[79] Adams AK, Wise-Draper TM, Wells SI, Human papillomavirus induced transformation in cervical and head and neck cancers, Cancers, 2014, 6, 1793–820.

[80] Broniarczyk J, Massimi P, Bergant M, Banks L, Human papillomavirus infectious entry and trafficking is a rapid process, J Virol, 2015, 89, 8727–32.

[81] Litwin TR, Clarke MA, Dean M, Wentzensen N, Somatic host cell alterations in HPV carcinogenesis, Viruses, 2017, 9.

[82] Sannigrahi MK, Singh V, Sharma R, Panda NK, Khullar M, Role of autophagy in head and neck cancer and therapeutic resistance, Oral Dis, 2015, 21, 283–91.

[83] Sannigrahi MK, Singh V, Sharma R, Panda NK, Radotra BD, Khullar M, Detection of active human papilloma virus-16 in head and neck cancers of Asian, North Indian patients, Oral Dis, 2015.

[84] Tsao SW, Tsang CM, To KF, Lo KW, The role of Epstein-Barr virus in epithelial malignancies, J Pathol, 2015, 235, 323–33.

[85] McKenzie J, El-Guindy A, Epstein-Barr virus lytic cycle reactivation, Curr Top Microbiol Immunol, 2015, 391, 237–61.

[86] Ressing ME, van Gent M, Gram AM, Hooykaas MJ, Piersma SJ, Wiertz EJ, Immune evasion by Epstein-Barr virus, Curr Top Microbiol Immunol, 2015, 391, 355–81.

[87] Nowalk A, Green M, Epstein-Barr virus, Microbiol Spectr, 2016, 4.

[88] Minhas V, Wood C, Epidemiology and transmission of Kaposi's sarcoma-associated herpesvirus, Viruses, 2014, 6, 4178–94.

[89] Avey D, Brewers B, Zhu F, Recent advances in the study of Kaposi's sarcoma-associated herpesvirus replication and pathogenesis, Virol Sin, 2015, 30, 130–45.

[90] Bhutani M, Polizzotto MN, Uldrick TS, Yarchoan R, Kaposi sarcoma-associated herpesvirus-associated malignancies: epidemiology, pathogenesis, and advances in treatment, Semin Oncol, 2015, 42, 223–46.

[91] Fooks AR, Cliquet F, Finke S, et al., Rabies, Nat Rev Dis Primers, 2017, 3, 17091.

[92] Hemachudha T, Ugolini G, Wacharapluesadee S, Sungkarat W, Shuangshoti S, Laothamatas J, Human rabies: neuropathogenesis, diagnosis, and management, Lancet Neurol, 2013, 12, 498–513.

[93] Schnell MJ, McGettigan JP, Wirblich C, Papaneri A, The cell biology of rabies virus: using stealth to reach the brain, Nat Rev Microbiol, 2010, 8, 51–61.

[94] Mackay IM, Arden KE, An opportunistic pathogen afforded ample opportunities: Middle East respiratory syndrome coronavirus, Viruses, 2017, 9.

[95] Gralinski LE, Baric RS, Molecular pathology of emerging coronavirus infections, J Pathol, 2015, 235, 185–95.

[96] Omrani AS, Al-Tawfiq JA, Memish ZA, Middle East respiratory syndrome coronavirus (MERS-CoV): animal to human interaction, Pathog Glob Health, 2015, 109, 354–62.

[97] Ilkow CS, Willows SD, Hobman TC, Rubella virus capsid protein: a small protein with big functions, Future Microbiol, 2010, 5, 571–84.

[98] Tyor W, Harrison T, Mumps and rubella, Handb Clin Neurol, 2014, 123, 591–600.

[99] Lievano F, Galea SA, Thornton M, et al., Measles, mumps, and rubella virus vaccine (M-M-RII): a review of 32 years of clinical and postmarketing experience, Vaccine, 2012, 30, 6918–26.

[100] Aguilar-Calvo P, Garcia C, Espinosa JC, Andreoletti O, Torres JM, Prion and prion-like diseases in animals, Virus Res, 2015, 207, 82–93.

[101] Fraser PE, Prions and prion-like proteins, J Biol Chem, 2014, 289, 19839–40.

[102] Wei X, Lawall C, Pepper P, Sporadic Creutzfeldt-Jakob disease in a Navy Commander: case report and review of the literature, Mil Med, 2015, 180, e174–76.

[103] Sharma S, Mukherjee M, Kedage V, Muttigi MS, Rao A, Rao S, Sporadic Creutzfeldt-Jakob disease – a review, Int J Neurosci, 2009, 119, 1981–94.

[104] Ironside JW, Variant Creutzfeldt-Jakob disease: an update, Folia Neuropathol, 2012, 50, 50–56.

[105] Thomas JG, Chenoweth CE, Sullivan SE, Iatrogenic Creutzfeldt-Jakob disease via surgical instruments, J Clin Neurosci, 2013, 20, 1207–12.

[106] Liberski PP, Sikorska B, Brown P, Kuru: the first prion disease, Adv Exp Med Biol, 2012, 724, 143–53.

[107] Montagna P, Gambetti P, Cortelli P, Lugaresi E, Familial and sporadic fatal insomnia, Lancet Neurol, 2003, 2, 167–76.

[108] Liberski PP, Gerstmann-Straussler-Scheinker disease, Adv Exp Med Biol, 2012, 724, 128–37.

[109] Collins S, McLean CA, Masters CL, Gerstmann-Straussler-Scheinker syndrome, fatal familial insomnia, and kuru: a review of these less common human transmissible spongiform encephalopathies, J Clin Neurosci, 2001, 8, 387–97.

Younis Ahmad Hajam, Rahul Datta, Sonika, Ajay Sharma,
Rajesh Kumar, Abhinay Thakur and Anil Kumar Sharma
Chapter 4
Parasitology

Abstract: The parasitology refers to interactions between two organisms of different sizes in which the benefitted one is known as the parasite and the other one which suffers is called the host. Although there are numerous prokaryotic and eukaryotic parasites, the main emphasis is based upon those which are responsible for most deadly illness leading to cause millions of deaths and billions of infections in humans every year. Public health agencies and research organizations largely hire parasitologists to test therapies, prophylactics and vaccines that prevent parasitic infections. Parasites can develop drug resistance, so it is therefore necessary to understand their genes, proteins, life cycles and evolution through research to monitor infections and to predict potential outbreaks.

Keywords: parasites, pathogenesis, diagnosis, parasitic infections

4.1 Pathogenesis of parasitic diseases

Parasitic diseases caused by eukaryotic organisms are of major concern and present global health issues. At present, 500 million people throughout the world are infected with malaria, schistosomiasis (>200 million) and hookworm or ascaris (close to a billion) [1]. Control and elimination of these diseases are also not completely successful due to various obstructions because these diseases affect mostly people in developing countries. In order to find a better solution to overcome these parasitic diseases, a clear understanding of pathogenesis of these diseases is the most important step. Pathogenesis is the process that how diseases develop and progress [2]. Protozoan and helminthes are responsible for many important human diseases. Several human diseases caused by protozoans are trypanosomiasis, toxoplasmosis,

Younis Ahmad Hajam, Sonika, Department of Zoology, Career Point University, Hamirpur 176041
Rahul Datta, Centre for Agricultural Research and Innovation, Guru Nanak Dev University, Amritsar 143005, India
Ajay Sharma, Department of Chemistry, Career Point University, Hamirpur 176041
Rajesh Kumar, Department of Zoology, Career Point University, Hamirpur 176041,
e-mail: drkumar83@rediffmail.com
Abhinay Thakur, Department of Zoology, DAV College Jalandhar, Punjab 144008, India,
e-mail: abhinay.thakur61@gmail.com
Anil Kumar Sharma, Department of Biotechnology, MMDU, Mullana, Ambala

https://doi.org/10.1515/9783110517736-004

leishmaniasis and malaria [3], whereas important helminth disease parasites include filariasis, schistosomiasis and cysticercosis [4].

Both protozoan and helminthes diseases have quite different pathogenesis. The mechanism of infection also varies according to the type of species as well as host–parasite adaptation and response. Several factors responsible for variability include parasite size, reproductive ability of parasite, type of injury, mode of infection, nutritional requirements, niche and most important is the parasite-derived proteases [5]. The damage caused by the parasites are mostly due to toxic secreations, mechanical pressure or either through seizing the nutrition of the host [6]. Parasitic species release toxic substances, viz. secretion, excretory products, etc., all of which somehow have toxic effects on the host. In America, Chagas disease, caused by the protozoan parasite *Trypanosoma cruzi*, is transmitted to humans through transfusion of blood (70%), congenital (26%), transplacental, organs (1%) (kidney and heart) as well as laboratory infection [7, 8]. It is a flagellate protozoan found in muscle cells, viz. smooth and cardiac muscle, and the infection resulting in inflammatory response [9, 10].

One of the most proposed pathogeneses of Chagas heart disease involves the appearance of *T. cruzi* in the heart or elsewhere, to propagate cardiac pathology [8]. During the infection process, mechanical damage occurs to the cell, tissue or organ of the host due to mechanical pressure from the parasite. For example, complement activation requires the presence of glycoproteins present on the surface of trypanosomes which resulted in the production of biologically active and toxic complement fragments and also release proteases and phospholipases when they lyse. These enzymes can produce host cell destruction, inflammatory responses and gross tissue pathology [11]. Various secretions, excrement released by parasites, stimulate the host cell or tissue to produce antibodies by acting as an antigen. However, when the same antigen stimulates the organism for the second time, pathological immune response known as the hypersensitivity reaction is produced, resulting in tissue damage and disorders [12]. Parasite completely depends upon the host for nutrition to carry out several biological processes, viz. growth, development and reproduction. Hookworm sucks the blood of the human by residing in small intestine before passing through your feces. Sucking of blood by hookworm takes place by biting the intestinal mucosa, which leads to blood loss within the intestine, anemia caused by iron deficiency and protein malnutrition [6].

4.2 Laboratory diagnosis of parasitic diseases

Diagnosis of parasitic diseases depends on various laboratory methods besides clinical history, symptoms, travel history as well as geographic location of patients [13]. During the last few decades, these methods remain inert. However, microscopy still plays a key role in laboratory testing for various blood and tissue parasites. Examination of blood smears relies on the microscope for detection and identification of several protozoan (*Plasmodium*, *Trypanosoma*, etc.) and the filarial nematode species (*Wuchereria*, *Brugia*, etc.) [14]. A major step in parasite detection and control is exact diagnosis. In order to diagnose the parasite, a patient's sample of feces, blood, sputum, biopsy specimens or skin scrapings is required, and appropriate tests must be assessed for diagnosis. Diagnosis of parasitic infection relies on collection and transport of the sample to the laboratory and the method of laboratory evaluation. However, several factors are taken into consideration during the diagnosis, viz. physiological relationship, age of the host, previous exposure to parasites, geographical location, previous use of anthelmintics and history of clinical diseases [15]. Recent development in technology has provided new diagnostic tools in parasitic detection. Important laboratory diagnostic approaches are:

1. Microscopy
2. Highly sensitive serology-based assays, viz. Falcon assay screening test (FAST)-enzyme-linked immunosorbent assay (ELISA) [16], dot-ELISA [17], rapid antigen detection system (RDTS) [18] and luciferase immunoprecipitation system (LIPS) [19].
3. Molecular techniques such as multilocus enzyme electrophoresis [20], nucleic assay-based diagnosis [13], real-time polymerase chain reaction (PCR) [21], loop-mediated isothermal amplification (LAMP) [22] and Luminex [23].

4.2.1 Microscopy

One of the major tools available for the diagnosis of parasites from blood smears, urine, tissue specimen, feces, lymph node aspirates, bone marrow and cerebrospinal fluid (CSF) is microscopy [24]. Most of the intestinal helminths and protozoan infections are detected through microscopy. However, other parasitic infections are diagnosed by synchronized microscopy with other procedures of diagnosis, including serology-based as well as molecular-based assays [13]. Protozoans found in blood circulation are diagnosed by preparing blood smear and stained with Romanowsky stain for the detection of babesiosis, theileriosis and trypanosomosis [24, 25], whereas blood smears stained with Giemsa or other appropriate stains are used for detection and identification of species of *Plasmodium*, *Babesia*, *Trypanosoma*, *Brugia*, *Mansonella* and *Wuchereria* [14]. Helminths are diagnosed by making a smear with saline and iodine, whereas in case of dysenteric specimen, smear should be made from the specimen containing blood and mucus without adding saline or stain and cover with the cover slip. Watch the slide under microscope to identify any larvae or helminthic eggs [26, 27].

4.2.2 Serologic assays

Serology plays an important role in diagnosis situation, where biologic samples or tissue specimens are unavailable. These assays are categorized into two parts, i.e., antigen-detection assays and antibody-detection assays [27]. These include ELISA, also called enzyme immunoassay (EIA), and all its derived tests such as the FAST-ELISA, dot-ELISA and LIPS. Other assays include the hemagglutination test, indirect or direct immunofluorescent antibody (IFA) tests, complement fixation test, and immunoblotting and rapid diagnostic tests [13]. These assays have both advantages and disadvantages. Disadvantage lies due to its invasive nature because the antibodies remain persistent after treatment and did not specify any active infection cross-reaction with other nematodes, particularly with filarial infections [28]. However, serologic assays are very easy to perform and are highly sensitive and specific as compared to other conventional parasitological tests. These assays show remarkable outcome in those patients that are asymptomatic or whose blood smears do not permit identification of the parasite [29, 30].

4.2.3 Falcon assay screening test-ELISA (FAST-ELISA)

In this assay, synthetic and recombinant peptides are used to evaluate antibody responses to an antigen [31]. Earlier, this assay was used for the study of malaria, fasciolosis, schistosomiasis and taeniasis [32–35]. Cross-reaction that occurs between closely related species (e.g., trypanosome) is the major drawback of this technique, which resulted in false results [24].

4.2.4 Dot-ELISA

The Dot-ELISA is a highly versatile technique for antibody or antigen detection that uses tiny amount of reagent dotted onto solid surfaces, viz. nitrocellulose and other paper membranes which strongly bind with proteins [36]. It is quick, simple to perform and interpret, low cost and moveable as well as reagent conservative which has been used widely for the detection of human parasitic diseases, viz. malaria, babesiosis, trypanosomiasis, cysticercosis, leishmaniasis and toxoplasmosis [36]. Dot-ELISA is a highly sensitive technique used in the detection of antineurofilament and antigalactocerebroside antibodies in the CSF of subjects infected with African trypanosomes [13].

4.2.5 Rapid antigen detection system (RDTS)

RDTS, also known as dipsticks, detects a specific antigen produced by malarial parasites by complexing them with captured antibodies embedded on a nitrocellulose strip [13, 37]. RDTS detects a parasite by color change on an absorbing nitrocellulose strip. Depending upon the type of antigen detected by RDTS, it can detect single species, i.e., *Plasmodium falciparum* by detecting histidine-rich protein-2 or parasite-specific lactate dehydrogenase, or other three species of malarial parasites infecting humans [37]. Introduced in the 1990s, RDTS relies on utilizing immunochromatographic lateral-flow-strip technology and can be applicable for diagnosis of malaria in hospitals as well as in remote health-care clinics [38]. This technique is used in African countries to overcome the misdiagnosis of malarial infections [39]. Specific factors are taken into consideration for introducing RDTS, and these are performance characteristics, operational characteristics and cost [37, 40].

4.2.6 Molecular approaches

4.2.6.1 Multilocus enzyme electrophoresis

Multilocus enzyme electrophoresis is a technique in which non-denatured protein (enzyme) migrates across a support medium (polyacrylamide, starch) under the influence of an electric field [20]. Due to their difference in net charges, shape and size, these proteins move at different rates on electrophoresis gel. For the last few decades, this technique has been used to identify various parasite organisms, viz. *Fasciola*, *Echinococcus* and *Trypanosoma* isolates [41–43]. Multilocus enzyme electrophoresis is one of the best cost-efficient techniques for exploring genetic phenomena at molecular level and proves to be very useful in systematic [44].

4.2.6.2 Nucleic assay-based diagnosis

Discovery of PCR has opened the way for parasitologists to diagnose the parasitic diseases in general as well as to compensate the drawbacks of microscopy and serology assays [13, 45]. Nucleic acid-based method provides more accuracy and specificity as compared to previous techniques and is very efficient in diagnosing even at low parasitized samples [46].

4.2.6.3 Real-time polymerase chain reaction (RT-PCR)

Real-time PCR, also known as quantitative-PCR (Q-PCR), is a procedure in which the quantity of a PCR product can be estimated in real time by the use of a fluorescent reporter [47, 48]. This technique is highly sensitive and proves to be useful in those parasitic diseases, where other laboratory diagnoses, viz. microscopy as well as serology, remain ineffective. Q-PCR is used in identification as well as quantification of parasite DNA in a closed system which decreases the risk of contamination in analysis of sample [49]. RT-PCR does not depend on any downstream examination such as electrophoresis or densitometry and is highly versatile, enabling multiple PCR targets to be assessed simultaneously [47]. Multiplexed RT-PCR uses several primer pairs for the detection of multiple pathogen in one sample [50]. Multiplexed RT-PCR is cost effective and it has rapid turnaround time [13]. Farcas et al. [51] reported that multiplexed RT-PCR proved helpful in differentiating drug-reactive strain of malaria. Multiplexed RT-PCR helps in detection of protozoan parasite [52].

4.2.6.4 Loop-mediated isothermal amplification (LAMP)

LAMP is effortless, efficient, fast and field-adaptable, has outstanding specificity and is an important technique that quickly amplifies the target DNA under isothermal conditions using a polymerase with strand displacement properties, usually the *Bacillus stearothermophilus* (Bst) polymerase. Amplification of DNA results in the formation of magnesium pyrophosphate precipitate used as an indication of positive reaction [53, 54]. LAMP has been used in detection of various human parasites, viz. *Trypanosoma* [53, 55, 56], *Plasmodium* [57], *Babesia* [58] and *Taenia* [59]. This technique is easy to perform and is carried out at constant temperature, and results can readily be assessed [13].

4.2.6.5 Luminex

Luminex is a bead-based xMAP (multianalyte profiling) technique that collaborates flow cytometry, fluorescent microspheres (beads), lasers and digital signal processing, and has the potential of concurrently measuring up to 100 different analytes in a single sample (http://www.luminexcorp.com/). Due to its extraordinary ability, Luminex could identify multiple organisms and different genotypes of one particular organism during the same reaction utilizing very low volume [13]. This approach may prove to be useful in the study of diagnosis of parasitic diseases [60]. Luminex is used in the detection of *Giardia duodenalis* and all species of *Plasmodium* [61, 62].

4.3 Intestinal and urogenital protozoa

Infection caused by protozoa is found throughout the world but prevalent mostly in countries where conditions of sanitation and hygiene are often low quality and so dispose to transmission. Major protozoan classes infecting human include flagellates, amoebae, ciliates and sporozoa, and based on their types, these may be commensals, pathogens or potential pathogens [63]. *Entamoeba histolytica*, *Giardia lamblia* (flagellate) and *Cryptosporidium hominis* (sporozoan) are the important human intestinal protozoan [64], whereas *Trichomonas vaginalis* (Flagellate) is the protozoan found in the urogenital tract [65].

4.3.1 Entamoeba histolytica

Entamoeba histolytica is an intestinal protozoan parasite belonging to the genus *Entamoeba*, which is found predominantly in humans and other primates. Around 50 million people throughout the world are infected with *E. histolytica* [64]. In humans, *E. histolytica* is the causative agent of amebic dysentery or amebiasis, dysentery syndrome (bloody, mucoid stools without fecal leukocytes).

4.3.1.1 Mode of infection

a) Intake of food or water containing cyst
b) Densely populated area where infection is more prevalent in individuals
c) Contaminated water plays a substantial role

4.3.1.2 Life cycle

Life cycle of *E. histolytica* encompasses an infective cyst form, metacystic trophozoite, invasive motile feeding trophozoite, precyst and metacyst stages [64]. Infection starts after ingestion of a mature cyst with four nuclei (10–16 mm) with contaminated hands, food or water. The young one inside the mature cyst is activated by a medium having pH 7 and above in the small intestine and splits from cyst wall and absorbs within the gut lumen where it divides by karyokinesis followed by cytokinesis resulting in eight uninucleate trophozoites. Trophozoites with the help of their pseudopodia move toward the large intestine where they divide by binary fission, feeding on the bacteria of the intestinal flora and on cell debris to produce cysts. Both stages are passed in the feces, cysts being most commonly found in formed stools, and trophozoites in diarrheal stools. In the intestine, dehydration of lumin triggers the condensation of trophozoites into a round mass (precyst), and a thin wall is secreted around the immature cyst. Cysts develop each with 1, 2 or 4 nuclei (metacyst) and are passed in the feces [66].

4.3.1.3 Symptom and treatment

A person suffering from amebiasis shows symptoms that vary widely and the symptoms depend upon the intensity and the location of the infection. Symptoms arise after the penetration of the epithelial cells in the gastrointestinal (GI) tract by the parasite. The patient suffers from severe diarrhea (blood and mucus in liquid feces), and within 1–2 weeks, frequent watery stools with blood and mucus accompanied by pain, tenesmus and fever [67]. Dysentery can last for months during which it may vary from mild to severe, resulting in weight loss and prostration. A person suffering with intestinal infections should be treated with diloxanide furoate, iodoquinol and paromomycin, which were found to be effective against intestinal lumen pathogens, whereas no such effect was seen against invasive disease [68]. For those patients with invasive disease, a combination of a luminal agent to eliminate parasites in the gut and an effective tissue amebicide, such as metronizadole or tinidazole, is required to prevent relapse [69]. For severe or refractory disease, dehydroemetine followed by iodoquinol, paromomycin or diloxanide furoate are suitable.

4.3.2 Giardia lamblia

Giardia lamblia is a flagellate protozoan parasite found in the small intestine of humans which is responsible for diarrheal illness known as giardiasis. The infection ranges from asymptomatic to prolonged illness with diarrhea, malabsorption and weight loss [70]. This parasite is very often found in children (6–10 years) as well as travelers from the endemic region. The infective stage is trophozoite, which is rounded at the posterior end, tapered posteriorly and flattened dorsoventrally [64].

4.3.2.1 Mode of infection

Drinking of contaminated water, e.g., sewage line.
Humans can get infected from sheep, beavers and dogs which act as a potential reservoir.
Ingestion of cyst-contaminated food or water or from direct hand-to-mouth contact.

4.3.2.2 Life cycle

The life cycle of *Giardia* consists of two different stages, i.e. the trophozoite that inhabits the intestine and the cyst (infective form) which is resistant to environmental condition. Infection starts with ingestion of contaminated food and water with infective form (cyst) [71–73]. In the stomach, excystation takes place due to the acidic environment and trophozoites are formed. The trophozoite is the infective stage and divides by longitudinal fission. It attaches themselves to epithelial cells of the intestine with the help of two ventral flagella. From digestive system, trophozoite migrates to encyst in the colon and cysts are passed out with feces. Life cycle is completed when the cyst comes in contact with a new host [71, 73].

4.3.2.3 Symptom and treatment

Infection resulted in diarrhea. Trophozoite attachment resulted in inflammation in the intestine as well as lesion on the mucosal cells [74]. Infected person unable to absorb vitamin B12, folate and carotene due to malabsorption. Diarrheic stools, weight loss, nausea, abdominal distension are other symptoms [75]. Metronidazole, tinidazole is mostly recommended by doctors.

4.3.3 Cryptosporidium hominis

Cryptosporidium hominis is a gut-dwelling intracellular parasite that infects both humans and animals and is a major cause of gastroenteritis (cryptosporidiosis) worldwide [76]. Sporozoite is the infective stage and transmission is usually by the fecal–oral route. It poses a serious concern for drinking water treatment because of the robust nature of the oocyst, which confers resistance to chlorine disinfection. Young children and patients with immune deficiencies are most prone to diarrheal illness due to infection caused by *Cryptosporidium* [77].

4.3.3.1 Mode of infection

Intake of oocysts during direct exposure with either human or animal excrement. Contaminated water and food acts as an indirect source of infection.

4.3.3.2 Life cycle

The life cycle of *Cryptosporidium* has been studied in detail [78, 79]. The life cycle starts with the ingestion of oocysts, and excystation in the small intestine resulted in four motile sporozoites. Sporozoite infects the epithelial cell of the intestine followed by asexual and sexual reproduction resulting in the formation of immature oocyst which sporulates inside the host. Mature oocysts formed by sporulation contain four infectious sporozoites that are shed in the feces. The newly emerged oocysts are thick walled and resistant to environmental condition [80]. Thin-walled oocysts initiate autoinfection within the host and are not detected in the host causing a prolonged disease [81].

4.3.3.3 Symptom and treatment

Most common symptoms include watery, sometimes mucoid, diarrhea, abdominal pain, low-grade fever, nausea and vomiting, leading to dehydration and weight loss. Incubation period ranges from 1 to 14 days. Patients with impaired cell-mediated immunity count $< 200/mm^3$ of CD4 T-cell result in severe, enfeebling and indeed life-threatening [82].

Treatment of water with secondary disinfection (UV or ozone): the multiple barrier concept is effective, which involves early protection of water source, improving water treatment mechanism along with proper water supply following safety measures such as regular washing of hands after using toilets and during caring of a person with dysentery [83]. 6-Carboxamide benzoxaborole AN7973 is a potent benzoxaborole drug candidate for treating cryptosporidiosis [84].

4.3.4 Trichomonas vaginalis

T. vaginalis is a pathogenic human-infecting trichomonad that inhabits female vagina and male urethra, epididymis as well as prostate gland, resulting in trichomoniasis. This is commonly transmitted through sexual intercourse [65]. It is comparatively larger in size as compared to other species. The trophozoite sometimes showed pseudopodia [64].

4.3.4.1 Mode of infection

Sexual intercourse
Damp washcloths
Newborns acquire infection while passing through the vagina

4.3.4.2 Life cycle

T. vaginalis required an optimum pH range, i.e., 5–6, to survive. It replicates by longitudinal binary fission. Due to the absence of cyst form, the organism does not survive well in external environment. *T. vaginalis* disturbs the pH of the vagina (4–4.5) maintained by a group of lactic acid-producing bacteria. Fluctuation in pH creates an ideal environment for the *T. vaginalis* to thrive there [64].

4.3.4.3 Symptom and treatment

The cell of the vaginal mucosa deteriorates due to infection resulted in inflammation and regular vaginitis followed by yellowish discharge leads to burning and itching. Swelling of prostate gland and urethritis occur in males [65]. Metronidazole and tinidazole are the most effective drugs used for treatment [64]. Regular douches of diluted vinegar are effective in maintaining vaginal pH.

4.4 Antiparasitic agents

Parasitic infections are responsible for millions of deaths every year and are most problematic in those areas of the world having limiting rate of economic development. Among various parasitic organisms, protozoan and helminth infections pose a serious threat to health care throughout the world and are responsible to cause significant health implications [85]. In order to treat infestations caused by these organisms, antiparasitic agents (Table 4.1) are mostly used, which are drugs used to treat various types of parasitic diseases, and various types of antiparasitic agents are discussed further.

4.4.1 Antiprotozoal agents

4.4.1.1 Antimalarial agents

Several advances have been made to provide effective treatment against malaria. Various derivatives of 4-aminoquinoline are mostly used for treatment. Chloroquine (4-aminoquinoline) is an antimalarial agent used for the treatment of malaria caused by all plasmodium species, viz. *P. vivax*, *P. malariae*, *P. ovale*, and *P. knowlesi* except *P. falciparum* [86], whereas amodiaquine is used against *P. falciparum*. Artemisinin is also one of the most effective and fast-acting antimalarial drugs available to date, extracted from the *Artemisia annua* [87].

4.4.1.2 Antiamebic agents

Infection caused by *Entamoeba histolytica* can be treated with antiamebic agents and these are divided into three type's viz. luminal, tissue, and mixed amebicides. Diiodohydroxyquinoline, diloxanide and paromomycin are luminal, whereas metronidazole and nitroimidazole are mixed amebicides used for the management of amebiasis [88]. Mixed amebicides are active in both tissue and intestinal lumen and are found to be most effective therapeutically. Tissue amebicides are rarely used.

4.4.1.3 Antigiardial agents

Metronidazole is the best drug used for management of giardiasis. However, treatment with either tinidazole, or quinacrine is recommended; other options are furazolidone and albendazole [64, 89].

4.4.1.4 Antitrichomoniasis agents

From the last few decades, metronidazole is the mostly used against trichomoniasis. It belongs to the drug family 5-nitroimidazole and has been reported to about 95% effective. Tindizole and seconidazole are also the treatment of choice [90]. In pregnant women, metronidazole is the safest drug to be recommended [65].

4.4.1.5 Antileishmanial agents

One of the best drugs for the treatment of leishmaniasis is sodium stibogluconate and meglumine antimoniate [91, 92]. Other alternatives available are meglumine antimoniate, pentamidine and amphotericin B [93].

4.4.1.6 Antitrypanosomal agents

Treatment of Chagas disease caused by *Trypanosoma cruzi* is done with the help of benznidazole and nifurtimox. Benznidazole is found to be effective up to 90% in acute cases [94]. However, pentamidine, eflornithine, melarsoprol and suramin are used for the treatment of diseases caused by *T. brucei gambiense* and *T. brucei rhodesiense* infection [95–98].

4.4.1.7 Antihelminthic agents

These agents are classified on the basis of infection of parasitic worm as well as the chemical structure of drugs.

4.4.1.8 Anticestodal drugs

Commonly used as a drug against cestodes (*Taenia saginata*, *Diphyllobothrium latum* and *Taenia solium*), infection is praziquantel and niclosamide [99, 100]. In case of hydatid disease and cysticercosis, albendazole is a highly recommended drug [89].

4.4.1.9 Antinematodal drugs

Praziquantel is one of the best drugs used for the treatment of infection caused by *Clonorchis sinensis*, *Schistosoma* and flukes, whereas albendazole is used to cure infections caused by roundworms, viz. ascariasis, trichuriasis, pinworm and hookworm infections [89].

Table 4.1: List of antiparasitic agents.

Parasite	Drug	Authors
Antimalarial drugs	4-Aminoquinoline, chloroquine, atovaquone, amodiaquine, artemisinin, OZ439, OZ277 (arterolane), artemisinins, imidazolopiperazines and imidazolopyrazines	Campbell and Soman-Faulkner [89], Srivastava et al. [101], Maser et al. [87] and Jacobs [102]
Antiamebic agents	Diiodohydroxyquinoline, diloxanide, paromomycin, metronidazole, nitroimidazole	Martínez-Palomo et al. [88]
Antigiardial agents	Metronidazole and quinacrine	Campbell and Soman-Faulkner [89]
Antitrichomoniasis agents	Metronidazole	Page [85] and Cudmore and Garber [90]
	Tindizole and seconidazole	Cudmore and Garber [90]
Antileishmanial	Sodium antimony gluconate (SAG), amphotericin B, miltefosine and paromomycin	Ghosh et al. [103]

Table 4.1 (continued)

Parasite	Drug	Authors
	Sodium stibogluconate and meglumine antimoniate	Rais et al. [91] and Yesilova et al. [92]
	Amphotericin B, miltefosin, paromomycin, paromomycin and sitamaquine	Bhargava and Singh [104]
Antitrypanosomal	Aromatic diamidines, trifluralin, allopurinol, eflornithine (DFMO, DL-α-difluoromethylornithine) and antimonials	Page [85]
	Benznidazole and nifurtimox	Rios et al. [94]
	Pentamidine, eflornithine, melarsoprol and suramin	Thomas et al. [95], Balfour et al. [96], Morgan et al. [97], Fairlamb and Horn [98]
	Fexinidazole and acoziborole	Croft et al. [105]
Anticestodal drugs	Niclosamide, disophenol and dichlorophen	Page [85]
	Praziquantel, niclosamide and albendazole	Cupit and Cunningham [99] and Chen et al. [100]
Antinematodal drugs	Emodepside, chlorpyrifos and piperazine	Page [85]
	Praziquantel and ivermectin	Cupit and Cunningham [99] and Liang et al. [106]

4.5 Blood and tissue protozoans of human

4.5.1 Trypanosoma

Trypanosoma is a multicellular parasitic protozoan with a complex life cycle. Collectively, *Trypanosoma brucei* is transmitted by arthropod vectors and to mammalian hosts [107, 108]. African trypanosomiasis is caused by the parasite species *T. brucei* with subspecies *T. b. gambiense* and *T. b. rhodesiense* transmitted through tsetse fly [108]. *T. gambiense* has three types of host: primary, reservoir and secondary or intermediate. Man is a primary host while reservoirs are domestic and wild animals. The intermediate host is tsetse fly *Glossina palpalis*. *T. cruzi* is responsible for American trypanosomiasis or Chagas disease which is spread through the *Triatoma* species of the Reduviidae family [109].

4.5.1.1 Mode of infection

The mode of infection is inoculative. The infected tsetse fly *G. palpalis* when bites a man in order to suck the blood, it transmits the infective stage parasite to the new definitive host. The meta-cyclic stage of the parasite along with the saliva of the fly reaches the subcutaneous blood of the host.

4.5.1.2 Life cycle

The vector tsetse fly, *Glossina*, carries the trypanosome within the midgut from the blood of host. After then moves to the salivary glands of fly whereby they can be transmitted during the next feeding. After inoculation within the host, the parasite can live freely within the bloodstream and evade mammalian host defenses through variable surface glycoproteins (VSG). The slender form secretes bloodstream stage-specific VSG to escape from the host immunity and proliferates [108]. As the parasite population increases, a morphologic stumpy form with division arrest occurs. It is in this stage it may be transmitted to another tsetse fly from the host. Within the new tsetse fly, the organism progresses to a procyclic form, whereby VSG is lost, and the organism is established again in the fly midgut. Cell division is again arrested, and they migrate to the salivary glands as epimastigote forms. These transition from another proliferative stage to a nonproliferative form, where they again reacquire VSG and are now capable to reinfect a new mammal at the next blood meal [108] (Figure 4.1).

4.5.1.3 Preventive measures

The following measures should be taken into consideration to control human trypanosomiasis:
1. Treat the infected human being with drugs under the advice of a physician.
2. The tsetse flies must be killed with suitable insecticides.
3. The bushes must be cleared along the streams near town or villages.
4. The reservoir host must be killed to control the infection.

4.5.1.4 Treatment and management

The treatment for all subspecies of *T. brucei* are different and depend on the stage of infection. The first stage of the disease caused by *T. brucei gambiense* is treated by pentamidine, i.e., first-line therapy. The route of administration is intramuscular for 1 week or intravenous with normal saline over 2 h. Furthermore, the treatment for the next stage disease includes melarsoprol and eflornithine. The latter one

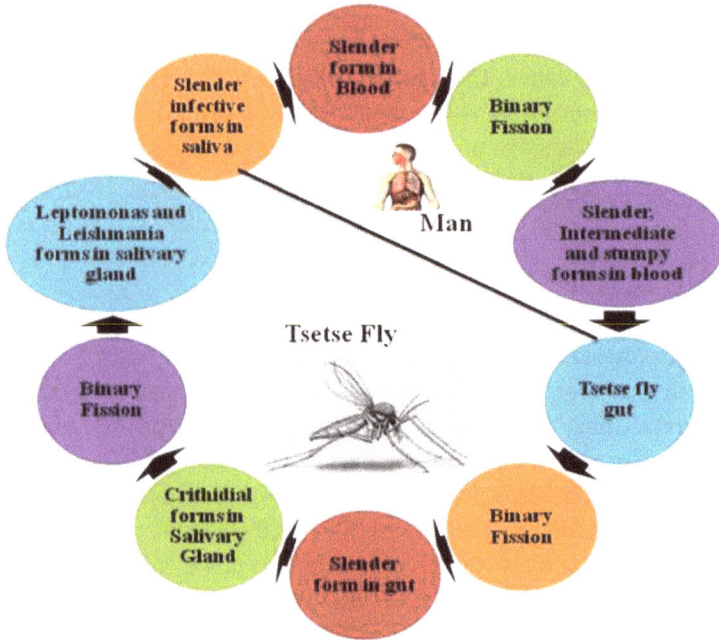

Figure 4.1: Diagrammatic representation of life cycle of *T. gambiense.*

has been found to be far better in lowering the mortality as compared to melarso-prol and is, therefore, the preferred medication for the second stage of disease [110]. Adverse reactions are similar to eflornithine and include pancytopenia, GI distress and convulsions. Similarly, suramin and melarsoprol are utilized for the first- and second-stage treatment of disease caused by *T. brucei rhodesiense*, respectively [107, 110].

4.5.2 Leishmania

Generally, humans are infected by three species of *Leishmania*, i.e., *L. braziliensis*, *L. tropica* and *L. donovani*. The disease caused by them is known as Leishmaniasis, which is found to be a major vector-borne [111, 112]. *L. donovani* causes visceral leishmaniasis, and other names are also available for the same, e.g., kala azar, dum dum fever and black fever. It is transmitted by female sand flies which belong to genera *Phlebotomus* and *Lutzomyia* [113]. Depending upon the area where these organisms occurs, they are called as old world species (Africa, Europe and Asia) and the new world species (America) [114].

4.5.2.1 Life cycle and mode of transmission

The life cycle includes one extracellular stage, which occurs within the body of sandfly and intracellular stage in humans. The two forms of parasites, i.e., amastigotes and promastigotes, exist in human and sandfly, respectively [115]. Initially, a parasitized sandfly (female) takes blood from humans. The promastigote form of parasite is converted into spherical, intracellular form of amastigote (no flagellum) by human macrophages. Furthermore, the multiplication of parasites occurs inside the macrophages by binary fission. These macrophages then show lysis. Similarly, other organs, e.g., bone marrow, the lymph nodes, spleen and liver, get infection [116–118]. The life cycle is demonstrated in Figure 4.2.

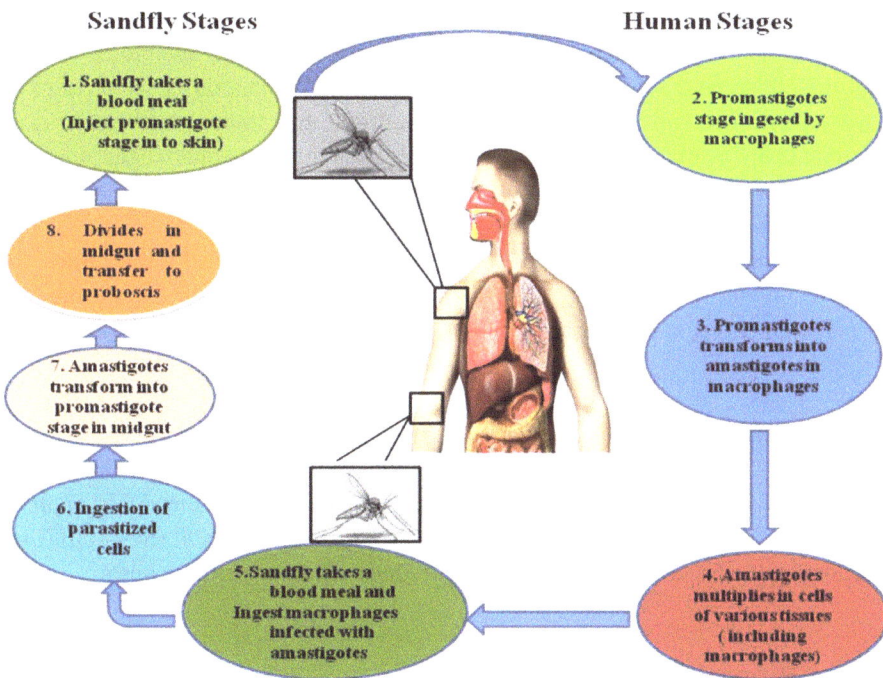

Sandfly Stages

Human Stages

1. Sandfly takes a blood meal (Inject promastigote stage in to skin)

2. Promastigotes stage ingesed by macrophages

8. Divides in midgut and transfer to proboscis

7. Amastigotes transform into promastigote stage in midgut

3. Promastigotes transforms into amastigotes in macrophages

6. Ingestion of parasitized cells

5. Sandfly takes a blood meal and Ingest macrophages infected with amastigotes

4. Amastigotes multiplies in cells of various tissues (including macrophages)

Figure 4.2: Life cycle of *Leishmania donovani*.

Due to hematophagous nature of female sandfly like mosquitoes, only female is responsible for spreading the infection [115]. These species of sandflies which are biological vectors mainly belong to the genera *Phlebotomus* and *Lutzomyia* [118].

4.5.2.2 Prevention measures

The destruction of vectors and their suitable habitat as well as their reservoir hosts should be done immediately. If humans get infections, then they may be treated with the help of suitable medicines.

4.5.2.3 Symptoms and treatment

Generally, the incubation period is 2–6 months, but can vary from 10 days to many years. Initially, the diseases spread gradually without showing any visible symptoms in patients [119]. The prolonged symptoms include undulant fever, loss of weight, reduced appetite, anemia and abdominal distension [112]. Additionally, dark skin coloration, coughing and chronic diarrhea are other symptoms too. Except this, in some cases of VL, post-kala-azar dermal leishmaniasis occurs after recovery [114]. The various types of oral, parenteral and topical medication drug treatments are available [120]. Since the 1940s, stibogluconate and meglumine antimoniate have been mainly utilized for treatment, but they are less in use due to their adverse side effects, resistance and cost. The regions where resistance is common, liposomal amphotericin B (LAMP) is found to be more effective. In future, the research on various new drugs, e.g., sitamaquine, paromycin and miltefosine may help in eradicating such diseases [119].

4.5.3 Plasmodium

The most widespread species causing malaria is *P. vivax*. It is assumed that approximately one-third of the world's population can be infected with *P. vivax* [121]. Usually, human and mosquito (female *Anopheles* mosquito) are its main hosts.

4.5.3.1 Lifecycle

The parasite is mainly transmitted in the form of sporozoites from the female mosquito to humans when it pierces the human body and sucks the blood for meal. Once the parasites enter into humans, they invade the liver cell known as hepatocytes and thereafter develop schizont which develops into merozoites in the blood after rupturing. This stage reveals malaria symptoms in the host body. The life cycle has been demonstrated in Figure 4.3.

Both species of *Plasmodium*, i.e. *P. vivax* and *P. ovale*, found to have dormant hypnozoite liver stage which may further cause relapse of infection when reentering into the main bloodstream [122]. The reticulocytes are mainly infected by merozoites

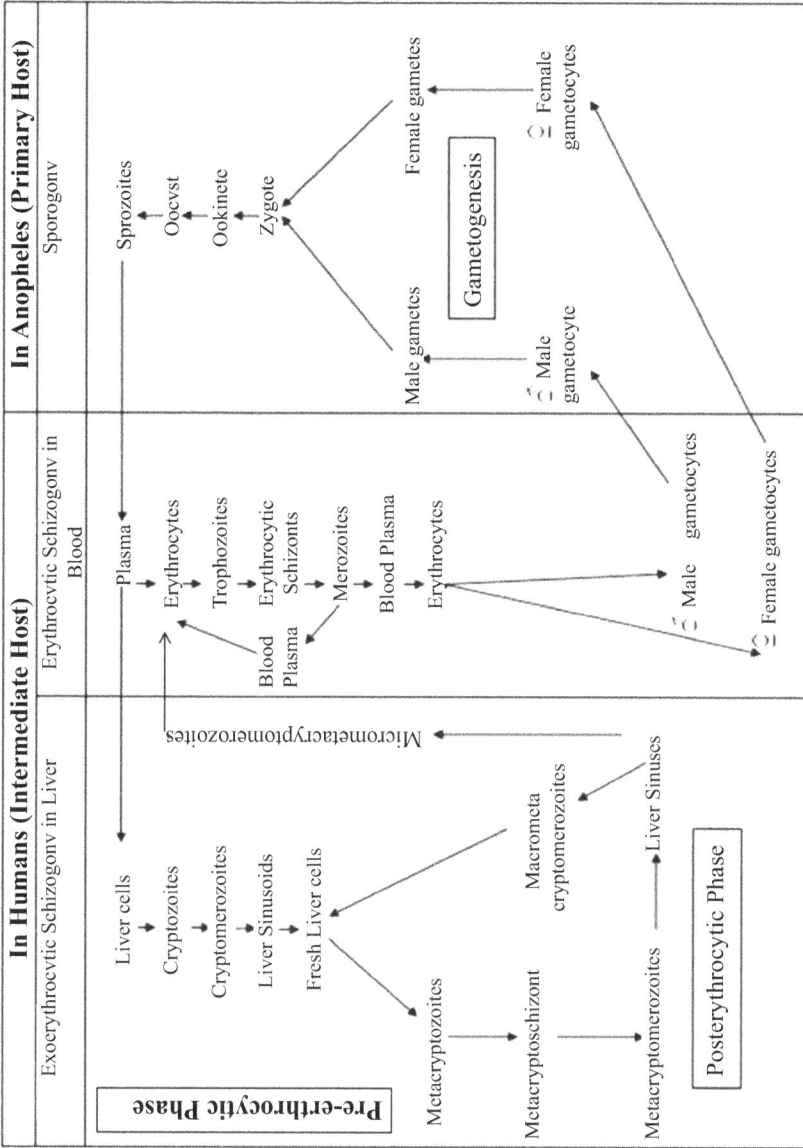

Figure 4.3: Graphical representation of *Plasmodium* life cycle.

of *P. vivax* unlike other species. This results in significant lower parasitemia levels in patients who are infected with *P. vivax* in comparison to those infected with *P. falciparum*. Although the level of parasitemia is low, but still it can result in significant disease due to increased host immune response [123]. The sexual and asexual multiplication of parasites takes place in humans. The life cycle of a parasite continues when a female mosquito again bites and inoculates its next human host.

4.5.3.2 Symptoms, treatment and management

4.5.3.2.1 Severe malaria
The symptoms of malaria including deep breathing and problem in respiration, prostration, anemia, fever, clinical jaundice and severe condition may involve dysfunctioning of various vital organs. Out of various drugs used, chloroquine is commonly used for treatment. In any case, if this drug is found to be resistant, then the alternate drugs atovaquone-proguanil, quinine plus either tetracycline or doxycycline or mefloquine were utilized. The use of intravenous artesunate is also recommended by the World Health Organization (WHO) in case of severe malaria, which is further based on medical research [124–126].

4.6 Helminthes

Helminthes that are commonly known as flat pathogenic worms refer to parasitic worms that are characterized by elongated, flat or round bodies and mainly include flukes and tapeworms. These may be flatworms or roundworms. The word "flatworms" has been derived from the Greek word meaning "flat," and roundworms are nematodes that means "thread" [127]. The helminthes have been divided as per the host, including "lung flukes," "extra-intestinal tapeworms" and "intestinal roundworms.". All helminthes that infect human beings have almost similar anatomical features. The morphological feature of general helminthes is the presence of suckers that differentiate it from other worms. The outermost layer of helminthes is known as cuticle or tegument which modifies into accessory sex organs in males [128].

The anatomy of helminthes body is very complex and requires keen observations to identify and locate internal organs, especially alimentary canal and excretory and reproductive systems. The tapeworms have a special character as they lack alimentary canal, and nutrients get absorbed and assimilated through the outer body surface. The blood flukes and nematodes are normally bisexual; however, all those which infect human beings are hermaphroditic [129].

4.6.1 General concept for the basis of classification

The helminthes are parasitic worms that can be separated and classified according to their morphological features, anatomy and the host inside which they reside. The body length of helminthes ranges from 1 mm to 1 m. All helminthes have well-developed organ system level of organization. The body of flatworms may be flattened or cylindrical covered with plasma membrane as in flatworms and with cuticle as in roundworms.

4.6.2 Classification

The helminths are broadly classified into three classes:
a) Nematodes
b) Trematodes
c) Cestodes

4.6.3 Nematodes (roundworms)

Nematodes that are commonly known as roundworms have typically 1,000 or less cells, thereby showing multicellularity. All nematodes appear worm-like, but are taxonomically as well as morphologically different from earthworms, wireworms or flatworms. The body of nematodes (roundworms) is divided into segments having elongated and cylindrical bodies; thus, it ranks the second largest phylum in the animal kingdom [129]. Nematodes derive their nutrition from microorganisms, plants, bacteriovores, fungivores, omnivores, predators and plant parasites [130]. *Ascaris lumbricoides* (Table 4.2), *Trichuris trichiura* (whipworm) (Table 4.3); *Ancylostoma duodenale* (Table 4.4) and *Necator americanus* (Table 4.5) (the two human hookworms); *Enterobius vermicularis* (pinworm); and *Strongyloides stercoralis* (Cochin-China diarrhea), *Dracunculus medinensis* (fiery serpents of the Israelites) and *Trichuris spiralis* live in cluster and are known as "soil-transmitted helminthes" because they are having some common phases in life cycles. "*E. vermicularis*" and "*T. trichiura*" are parasites on intestines, whereas others complete their life cycle in both intestine and tissues. Members of nematodes are discussed further.

4.6.4 Trematoda

Trematoda is another class of Platyhelminthes that causes infections worldwide. Trematodes are generally known as flukes due to the presence of conspicuous suckers. Suckers act as organs of attachment to the host to extract nutrition [142]. These

Table 4.2: Life cycle of *Ascaris lumbricoides*.

Ascaris lumbricoides	Description	Epidemiology	Diagnosis	Life cycle	Symptoms	Treatments
	A. lumbricoides is the largest of all nematodes that infect human beings. It has been estimated that nearly 1,300 million individuals all over the world have been infected with this parasite, leading total morbidity to about 120–220 million which stands 8–15% of the total infected people [131].	There are about 1 billion infected cases due to ascaris, out of which 20,000 deaths occur every year. The infection of *Ascaris* may occur at any stage of life; however, reports show that the infection is more severe in children of 5–9 years. Severity of infection was found more in rural population in comparison to urban one.	The diagnosis of ascariasis and related infection is possible only after examination of stool of the infected person for the presence of eggs of the parasite.	Ingestion of contaminated food has infected eggs. These infected eggs hatch into larva in the upper region of small intestines. The larvae enter into the lymphatics. The larvae then migrate into the liver, heart and lungs and they migrate into the bronchi, then pass on down to esophagus and from there goes down to small intestine and finally mature over there.	Abdominal pain. Loss of weight, anorexia, bloated abdomen, loose stool and vomiting in some occasional cases, cough, wheezing, dyspnea and substernal discomfort.	Albendazole, mebendazole, levamisole and pyrantel pamoate

Table 4.3: Life cycle of *Trichinella spiralis*.

Trichinella spiralis	Description	Epidemiology	Diagnosis	Life cycle	Symptoms	Treatments
	Trichinella is distributed around the world. It causes a disease known as trichinellosis [132]. Consumption of bloody or unprocessed meat (pork). Besides human beings, other mammals like wild carnivores and horses are sources of infection. *Trichinella* comprises nine species [133].	Every year about 10,000 cases, who are suffering from trichinellosis are reported [132]. The most of the cases are those who are consuming infected/ raw meat from common animals [134].	Detection methods include serum examination and muscle biopsy. For the examination of "leukocytosis", "eosinophilia", to detect the antibodies [132].	Infective larvae are ingested by mammals along with contaminated food and pass on to the small intestine region. After invading the epithelial wall of small intestine, larvae molt four times to attain sexual maturity. After copulation between the male and female parasites, the females begin to release newly born larvae in about a week time after infection.	Abdominal pain. Loss in weight, anorexia, swollen abdomen, intermittent loose stool and vomiting in some of the cases.	Albendazole 500 mg 2 times in a day for 10–14 days (or) Higher dose is 400–500 mg thrice in a can taken for next 10 days. "World Health Organization" recommends antihelminthic medications (mebendazole, albendazole, pyrantel or levamisole) for pregnant women later than in their first trimester [135].

Table 4.4: Life cycle of *Enterobius vermicularis*.

Enterobius vermicularis	Description	Epidemiology	Diagnosis	Life cycle	Symptoms	Treatments
	Enterobius vermicularis is the most common nematode that causes infection all over the world. Only natural hosts are humans for this infection.	The rate of infection varies from male to female 2:1. It generally negatively impacts children having age less than 18 years. Mode of infection is in contact with unhygienic clothes, bedding, etc. The most common transmission mode is fecal–oral.	Detection approach includes cellophane tape test or pinworm paddle test.	During night, female adult worms and ova enters into the anal area and lays numerous eggs in the perianal area. This egg laying causes itching and pruritus. Hatching of eggs takes place near the anal area causing itching and scratching, and this causes perianal pruritus.	Dysentery and vomiting	Albendazole: on empty stomach, 400 mg/one-time dose. Mebendazole: single dose (100 mg). Pyrantel pamoate: dose of 11 mg/kg.

Table 4.5: Life cycle of *Ancylostoma duodenale*.

Ancylostoma duodenale	Description	Epidemiology	Diagnosis	Life cycle	Symptoms	Treatments
	Ancylostoma duodenale are blood feeding and have been reported worldwide, especially among people in tropical and subtropical countries with low socioeconomic status [136, 137].	The parasite infects children, mainly infants aged less than 6 months [138]. The prevalence of infection rises with the increase in age [139]. The infections have been reported in the middle-aged people and sometimes even in the people above 60 years of age [140].	As far as diagnosis is concerned, not any standard investigative test available for assessing infection. Examination is generally delayed among majority of population as patients report nonspecific gastrointestinal complaints. Nowadays, blood and fresh stool samples are examined for diagnosing the infection.	Eggs from fecal matter penetrates the skin of feet and may enter the small intestine.	Gastrointestinal complaints	Benzimidazole, levamisole, mebendazole and pyrantel pamoate

Table 4.6: Life cycle of *Dracunculus medinensis*.

Dracunculus medinensis	Description	Epidemiology	Diagnosis	Life cycle	Symptoms	Treatments
	It is a parasite of human beings mainly reported in North, West and Central Africa, southwestern region of Asia, northeast of North America and China. The adult worm *Dracunculus medinensis* is shaped like a rope, cylindrical. Female worms are 500–1,200 × 0.9–17 mm, 12–18 months old, while male worms are 12–29 × 0.4 mm.	The epidemiology of dracunculiasis reports that a paratenic host is associated with the life cycle of a parasite. It may be an animal inhabitant to aquatic life, where it feeds upon copepods and other planktons. The infection harbors and subsequently transmitted to a definitive host.	The infection can be examined by assessing the egg and larvae in fecal matter [141].	Copepods enter the human body through the contaminated water. After the death of copepods, larvae penetrate the host's stomach, mature and reproduce.	Nausea, itching, rash, blisters on skin, diarrhea, dizziness, localized edema, reddish	Clean water, hygiene

are endoparasites mainly of mollusks and vertebrates. Majority of trematodes requires two or more hosts (primary and secondary) to complete their complex life cycle. The primary hosts are generally the vertebrates inside which flukes undergo sexual reproduction. The intermediate host is required to complete the reproduction process and is usually a snail. The trematodes or flukes include nearly 18,000–24,000 species, divided into two subclasses [143]. They are obligatory parasites of mollusks but may also infect turtles and fishes.

The class Trematoda may be divided into two subclasses:
I.   Monogenea
II. Digenea

On the basis of targets, trematodes may be divided into the following groups:
1. **Intestinal:** *Fasciolopsis buski*, *Heterophyes heterophyes*, *Echinostoma ilocanum* and *Metagonimus yokogawai*
2. **Liver/lung:** *Clonorchis sinensis*, *Opisthorchis viverrini*, *Fasciola hepatica* and *Paragonimus westermani*
3. **Blood:** *Schistosoma mansoni*, *Schistosoma haematobium* and *Schistosoma japonicum*

The description of few of them is given further.

4.6.5 Cestoda

Cestoda, or tapeworm that belongs to phylum Platyhelminthes, is a parasitic flatworm. It acts as endoparasites and inhabits the alimentary canal of various vertebrate animals, including humans. It is a flat, parasitic, hermaphroditic tapeworm with complex life cycles. The larvae of this parasite may infect both invertebrates and vertebrates. The life cycle requires more than one intermediate host to absolute its life cycle; in each intermediate host tapeworm completes a specific developmental phase. The body of the adult cestode is chiefly divided into scolex, neck and strobila regions. The scolex is located on the anteriormost end of worm. The neck which is unsegmented is situated posterior to the scolex and is the narrowest part of the worm. Proglottids differentiate from the neck and newly formed proglottids push the older ones that generate the chain of proglottids later termed the strobila. During this time, the worm also matures sexually. However, the more posteriorly located proglottids are more matured and developed. The maturity of reproductive systems allows a free subdivision of the strobila into immature, mature and gravid proglottids. The reproductive organs are evident in immature proglottids but are less functional, whereas they are fully functional in mature proglottids. Few of the cestodes are *Taenia solium*, *Taenia saginata*, *Diphyllobothrium*, *Taenia Hymenolepis nana*, *Hymenolepis diminuta*, *Echinococcus granulosus*, *Echinococcus multilocularis* and *Spirometra* spp. Description of few of them is given further.

Table 4.7: Life cycle of *Fasciolopsis buski*.

Fasciolopsis buski	Description	Epidemiology	Diagnosis	Life cycle	Symptoms	Treatments
	It is a largest parasite of an intestine that infects human beings. This worm is prevalent to Asia, specifically occurring in Taiwan, Thailand, Laos, China, Bangladesh, India and Vietnam.	Disease is prevalent in Southeast Asia. This parasite infects usually the host by intake of encysted metacercariae from clean plants that are fertilized by human or swine excreta.	Diagnosis may be done by fecal examination for the presence of eggs. Sometimes whole worms may be retrieved through vomiting or feces of infected persons.	Eggs become embryonated and develop into miracidia that enters snail, which act as an intermediate host. There it develops into cercariae; it encysts as metacercariae on available water- dwelling plants. The definitive host is usually a mammal that gets infected by ingesting metacercariae. After ingestion by the mammalian host, metacercariae encysts in its duodenum, then attaches on the intestinal wall and then changes into adults in about 3 months.	Microcytic anemia. Cardiac, thyroid and renal dysfunctions normally get affected. Ultrasound of abdomen may reveal mild ascites.	Praziquantel, metronidazole and albendazole

Table 4.8: Life cycle of *Heterophyes heterophyes*.

Heterophyes heterophyes	Description	Epidemiology	Diagnosis	Life cycle	Symptoms	Treatments
	It is a parasitic worm that mainly infects humans who used to eat undercooked aquatic animals, especially fishes infected with the metacercariae.	The infection is most commonly seen in fishermen of those who live in coastal areas.	Infection usually remains asymptomatic and mild.	Adults release embryonated eggs, which are ingested by the snail. After hatching, these eggs release miracidia larva that penetrates through the intestine of snail. Miracidia undertakes several developmental stages inside the body of snail and finally metacercariae is released. This enters the body of fishes and to the final host by intake of uncooked fish carrying metacercariae.	Discomfort in abdomen, loss of appetite, chronic diarrhea with mucous and nausea. In some cases, inflammatory reactions may occur.	Praziquantel 25 mg/kg. Triclabendazole 10 mg/kg is the first-line therapy.

Table 4.9: Life cycle of *Echinostoma ilocanum*.

Echinostoma ilocanum	Description	Epidemiology	Diagnosis	Life cycle	Symptoms	Treatments
	Echinostoma parasites are the most harmful parasitic worms among all helminthes that infect human beings and other animals.	Infection is mostly found in Asian continent, where practices like ingestion of undercooked fish and other animals or drinking of impure water are prominent [144].	Diagnosis can be done through careful examination of eggs. Adult fluke may sometimes be recovered for confirming the infection. Endoscopy may also be carried out in severe symptoms.	Unembryonated eggs are released in fecal matter of the infected animal. It develops into miracidia after about 3 weeks. It further develops into cercariae that may encyst as metacercariae. Infection spreads through metacercariae to secondary hosts.	Severe gastrointestinal symptoms, including epigastric or abdominal pain accompanied by diarrhea, easy fatigue and malnutrition	Praziquantel 25 mg/kg

Table 4.10: Life cycle of *Metagonimus yokogawai*.

Metagonimus yokogawai	Description	Epidemiology	Diagnosis	Life cycle	Symptoms	Treatments
	Metagonimus is an important helminth that causes infection in common public [145]. It is abundant in southern and eastern coasts [145].	The infection caused by the parasite is endemic or specifically prevalent in "Japan, Korea, China, Taiwan, the Balkans, Spain, Indonesia, the Philippines and Russia."	The diagnosed can be made by examining eggs in fecal matter. Infection may be confirmed if adult worms are found in feces. In certain cases, ELISA may be performed.	Adults release fully embryonated eggs, which are ingested by snail (primary host). The eggs hatch into miracidia that enters the intestine of snail. There it develops into cercariae and finally develops into metacerceriae. The final host gets infected by consumption of unprocessed fish carrying metacercariae.	Gastrointestinal problems	Praziquantel

Table 4.11: Life cycle of *Clonorchis sinensis*.

Clonorchis sinensis	Description	Epidemiology	Diagnosis	Life cycle	Symptoms	Treatments
	This parasitic helminth is endemic to Asian continent.	*C. sinensis* is a very prevalent worm parasite. Nearly 15 million citizens get unhygienic every year, and about 200 million people are at constant risk of infection [146–148].	Clonorchiasis has very nonspecific symptoms and it goes misdiagnosed most of the time. Diagnosis may be carried out by examining stool and blood of infected people.	Host discharges eggs in feces and then eggs enter the body of snails that act as primary hosts. Eggs hatched into miracidia undergo different development stages. Cercariae stage is swallowed by snails. After maturity, cercariae are shed in water from the body of snail. From there, it enters the second host, usually freshwater fish and enters definitive host when raw or undercooked fishes are eaten.	Fatigue, nausea, stomachache, jaundice and in severe cases hepatosplenomegaly may occur.	Praziquantel 25 mg/kg thrice daily [149]

Table 4.12: Life cycle of *Fasciola hepatica*.

Fasciola hepatica	Description	Epidemiology	Diagnosis	Life cycle	Symptoms	Treatments
	Fasciola is highly endemic in some regions of the developing world. As per recent reports, more than 2.6 million people get infected by this parasite every year all over the world [137].	As per WHO, fascioliasis is a zoonosis foodborne trematode disease. It causes loss to livestock, especially sheep and cattle. This disease is most prominent in developing countries of the world.	It is tough to diagnose fascioliasis due to very uncommon/unspecific symptoms [150]. Serology proves to be useful in diagnosing the infection.	The immature eggs are released through the stool and gets embryonated in freshwater in about 2 weeks. Eggs develop into miracidia which enters an intermediate host, usually snail. After undergoing various stages, metacerceriae are discharged in water. It invades the intestinal wall. Flukes emerge out and enter biliary ducts through the liver parenchyma. The parasite gets mature into adults and starts producing eggs. It may take more than 3–4 months to get finally mature and cause infection in human beings.	Biliary pain, fever, jaundice, unspecific pain in gastrointestinal region	Albendazole and metronidazole

Table 4.13: Life cycle of *Paragonimus*.

Paragonimus	Description	Epidemiology	Diagnosis	Life cycle	Symptoms	Treatments
	It is also known as prevalent hemoptysis, oriental lung fluke infection and a major foodborne zoonosis contagion.	It is a parasitic disease where human beings may act as definitive hosts. The disease is endemic to Asian and African continents. There are about 50 species of *Paragonimus* out of which about 11 are known to cause infections in humans [151].	The ova/adult parasites can be seen in stool. Diagnosis may also be carried out through blood test. Adult worm may be detected in the excised tissue or cyst or at necropsy or pleural aspirate [151] or expectorated in the sputum.	Eggs are discharged in unembryonated form in the sputum, or passed with stool. Eggs hatch into miracidia and invade first the intermediate host. It develops inside the body of snail and cercariae stage invades the second host, usually crab or crayfish. There they develop into metacercariae. After traveling through the digestive tract, it enters into the lungs, where they finally mature into adults.	Gastrointestinal problems	Praziquantel 25 mg/kg

Table 4.14: Life cycle of *Schistosoma*.

Schistosoma	Description	Epidemiology	Diagnosis	Life cycle	Symptoms	Treatments
	Schistosoma species infects humans being in Africa and also Middle East and South America.	Schistosomiasis used to infect more than 250 million people every year.	The infection can be diagnosed by identifying eggs in stools, and antibody detection through ELISA	Eggs hatched to release miracidia penetrates into snail. It develops into sporocyst that further develops into cerceriae. It grows into schistosomulae that enter into liver and mature into adult.	Stomach pain, loose stool	Albendazole and metronidazole

Table 4.15: Life cycle of *Taenia solium*.

Taenia solium	Description	Epidemiology	Diagnosis	Life cycle	Symptoms	Treatments
	Taenia solium is distributed among the developing countries, especially in the rural communities, where man upholds close contact with pig, cattle and different domestic animals.	Taeniasis infectious disease is found observed in the whole world with incompatible degree of occurrence [152]. The disease is associated with animals and humans playing a great role in its transmission and epidemiology of taeniasis and cysticercosis [153, 154].	Diagnosis is carried out by examination of stool.	Life cycle begins with the infection in cattle and pigs due to egg-contaminated vegetation. After hatching, eggs invades the intestinal wall through which they enter striated muscles. Cysticerus is the infective stage that infects human beings by the ingestion of undercooked/infected meat.	Abdominal pain, fever	Praziquantel 5–10 mg/kg

Table 4.16: Life cycle of *Echinococcus granulosus*.

Echinococcus granulosus Description	Epidemiology	Diagnosis	Life cycle	Symptoms	Treatments
It belongs to Taeniidae family. Six different types of *Echinococcus* have been reported, out of which four species are of significant importance. It infects animals like dogs. It requires two hosts to complete its life cycle [155].	This parasite is widespread in "South America," "Eastern Europe," "Russia," "the Middle East," and "China." Most of the domestic animals including sheep, goats, camels, horses and cattle act as intermediate hosts [158]. In our country (India), Andhra Pradesh and Tamil Nadu are the most affected states [157].	The diagnosis may be carried out by serological examination.	Gravid proglottids are discharged through fecal matter that causes infection. The eggs are then engulfed by a suitable host. Hatching results into release of oncospheres that invades in the intestinal wall and enters in the liver as well as lungs. There they turn up into hydatid cyst. Finally, they infect definitive host through injection of contaminated food.	Abdominal pain	Mebendazole and albendazole

Table 4.17: Life cycle of *Echinococcus multilocularis*.

Echinococcus multilocularis	Description	Epidemiology	Diagnosis	Life cycle	Symptoms	Treatments
	It is a small cyclophyllid tapeworm present broadly in the Northern Hemisphere.	It is prevalent in European countries and infects the population residing in close proximity of domestic animals. The infection is also found in regions of central Europe, much of Russia, the "Central Asian republics" and "western China," "the northwestern parts of Canada" and "western Alaska" [158].	Radiography	The parasite requires specific hosts like foxes and rodents [161]. The proglottids discharge eggs through the feces. These are engulfed by a primary host like sheep, goat, cattle and horses. The eggs are hatched into oncospheres that turn into a hydatid cyst. It enlarges into protoscolices that finally attach to the intestinal walls of the definite host [160].	Abdominal pain	Mebendazole. In occasional cases, surgery may be required.

4.7 Arthropoda

The body of arthropods is covered with exoskeleton, divided into the number of segments, appendages are paired and joined, and symmetry is bilateral. A distinguished feature of arthropods is joined appendages covered with chitin and in some members it is mineralized by calcium carbonate. Rigid cuticle suppresses the growth, which is the reason arthropods replace their exoskeleton periodically by molting. Arthropods constitute phylum Euarthropoda, which constitutes insects, arachnids, myriapods and crustaceans. Arthropod includes phylum Euarthropoda and phylum Onychophora. Several species are having wings to adopt them under various environmental conditions. This facilitates arthropods to become the largest species-rich and adopts under all ecological conditions in mainly environments. Arthropods constitute 80% or more than that among the existing animal species. They are microscopic such as crustacean *Stygotantulus* up to the Japanese spider crab. Arthropods are medically important such as vespids, ticks, mosquitoes, mites, flies and fleas. These are diseases spreading illness and also connected through the T lethal allergic reactions, and generate probable toxins. In the present era, Arthropod-borne diseases are causing more threat to humans and animals, and are now the main reason for fatality and morbidity to the whole humankind. The prominent arthropod caused diseases include malaria, and numerous deaths occur per year. "Dengue," "sleeping sickness," "leishmaniasis," "Chagas' disease" and "X" are imperative intimidation in the world. Arthropods spread their diseases by various modes based on the type of vector and organism.

4.7.1 Vector-borne diseases

Vector-borne infections are diseases that get transferred when infected arthropod species (mosquitoes, ticks, triatomine bugs, sandflies and blackflies) bite. These vectors are cold-blooded (susceptible toward weather conditions). Typical weather conditions affect the continued existence and reproduction rate in vector, which directly affects the habitat fitness, allocation and abundance; strength and sequential approaches of vector activity (especially rate of biting) all over the year; and rates of growth, continued existence and duplication of microorganisms inside the vectors [161].

4.7.1.1 West Nile fever

The causative agent for West Nile fever is West Nile virus (WNV), which belongs to the Flaviviridae family and resembles with Japanese encephalitis antigenic group. When infected *Culex* spp. bite the birds and rarely humans, it leads to West Nile fever. In "European countries," the virus has been detected in "mosquitoes," "wild rodents," "migratory birds," "hard ticks," "horses" and "human beings." The prevalence of *Culex modestus* vector is associated with "temperature, humidity, rainfall and sunshine" as well. These variables imperatively elevate throughout the epidemic phase [162]. In 1996–1997, outbreak occurs in southeastern Romania and in 2000 in Israel.

4.7.1.1.1 Diagnostics

Infection of WNV is examined mainly on medical basis and responses of antibody [163], which includes the occurrence of anti-WNV immunoglobulin (Ig)M mostly in CSF [164]. In humans, replication of WNV occurs in monocytes and often elevated potency in "polymorphonuclear leukocytes" which causes transmission through transfusion of blood [167]. Therefore, numerous fast diagnostic methods have been designed for blood donor test such as nucleic acid testing, an "amplified transcription technique" for the detection of WNV-infected individuals [166].

4.7.1.1.2 Treatments

Present therapeutic alternatives against WNV are mostly used as supportive. Four USDA-licensed vaccines are accessible for horses, which includes "two inactivated WNV," "one nonreplicating live canary pox recombinant vector vaccine" and " one inactivated flavivirus chimeric vaccine." In some conditions, "passive immunization" is given to patients; however, "passive immunization" is associated with some adverse effects such as unintentional transmission of "blood-borne pathogens" and does not depend on the donor's antisera quality, cost and allergic situations.

4.7.1.2 Dengue

In India, dengue is a major health issue; however, in the last decade, serious new cases of dengue have been reported in different cities of India. Dengue influences each and every community of the society; however, poor people are highly affected. The WHO evidenced that dengue fever infects 50–100 million persons in each year. The features of typical "dengue fever" or "breakbone fever" are sensitive commencement of towering body temperature for 3–14 days prior to biting by the infected mosquito. Symptoms range from pain in frontal, pain in retro-orbital, "myalgias," "arthralgias," "hemorrhagic manifestations," "rash," and anorexia and nausea. It is transmitted as arboviral human disease due to the supply of water via pipes. *A. aegypti* (also known as "yellow fever mosquito") survives in urban environments, which occurs by traveling.

4.7.1.2.1 Epidemiology

Epidemiologic studies reported that temperature is one of the most important for the transmission of dengue in urban regions. Change in climate projection depends on the humidity for 2,085, which revealed that spreading of dengue moves the latitudinal [167]. In moderate regions, weather change may lead to more elevation in the duration of the spreading period [167]. The consequence of increase in optimum temperature affects the spreading of dengue in "southern Europe." The eruption of dengue virus (DENV) takes place in "Asia" and "America," and it has been evidenced that dengue epidemiology elsewhere. In "Papua New Guinea," 52% to 57% malaria in feverish patients and dengue fever was found in 8%; 83% is having dengue IgG or having the risk of high DENV. Medical symptoms include "fever, vomiting, headache, abdominal pain, hepatomegaly, myalgia, bleeding manifestations, generalized weakness, cough, splenomegaly, rashes and diarrhea."

4.7.1.2.2 Treatment

No specific treatment is available for WNV infections [168], although various screening activities are available to identify possible suppressors of "WNV" contagion such as "7th EU Framework grant agreement number 260,644." WNED patients classically require intensive-care unit-related interference and life sustains therapy, and then complete the treatment for complcte recovery [168]. Various Igs along with essential titers of antibodies are used against WNV in some specific cases of infection via SOTs [169].

4.7.1.3 Chikungunya fever

Chikungunya virus (CHIKV) is transmitted by *Aedes* (mosquito *Aedes*) and transmitted to human being by "*A. aegypti*" and "*A. albopictus*." The causative agents of "chikungunya fever" by a virus, which belong to the genus Alphavirus and family Togaviridae. In 2007, the attack of chikungunya fever occurred in northeastern Italy, and first cases were observed in European continent [170]. Huge number of *A. albopictus* mosquitoes were found in traps in the areas where cases are Vector. However, *A. albopictus* and CHIKV were found in Italy accidentally; *A. albopictus* in Europe depends on various climatic factors, including mild winters, mean yearly rainfall more than 50 cm and mean summer temperatures greater than 20 °C [171]. Number of vectors and considerable factors having epidemic potential are associated with time and seasonal activity.

4.7.1.3.1 Epidemiology

"*A. aegypti*" (*Stegomya aegypti*) and "*A. albopictus*" (*Stegomya albopictus*) are main causative vectors for CHIKV. *A. aegypti* is the primary vector for the transmission of CHIKV, and unpredictably it was identified as a second chief viral vector in both the places: in Reunion island *A. Aegypti* is considered as rare and in Madagascar, India and Gabon, mosquito species are prevalent. This eruption of *A. albopictus* was found as the main vector of CHIKV. However generally due to point alteration in the "E1 protein of CHIKV" increases viral uptake, duplication and spreading by the vector. Later on, *A. albopictus* has become able to acclimatize to the peridomestic environment [172].

In Africa, various types of other mosquitoes are also associated for the spreading of CHIKV, viz. "A. spp." (*A. furcifer, A. taylori, A. vittatus, A. fulgens, A. luteocephalus, A. dalzieli, A. vigilax, A. camptorhyntites* and *A. africanus*) and also *Culex annulirostris* and *Mansonia uniformis* [173].

4.7.1.3.2 Diagnosis

Differential examination of chikungunya fever differs from the serological study. It has been observed that fever mutually through polyarthralgia having "84% sensitivity with a 74% positive predictive value (PPV)" and an "83% negative predictive value (NPV)" [174]. Current eruption of chikungunya fever in Gabon, fever and arthralgia having 73% sensitivity for PPV and NPV of 79% and 44%, respectively [175]. Staikowsky et al. [178] studied that CHIKV patients had higher increase in "skin itchiness, arthralgia on feet and wrist, asthenia," increased temperature and a slowdown incidence of digestive indicators and pruritus than CHIKV-negative patients [176]. The symptoms of dengue and chikungunya fever differ largely, viz. in chikungunya "arthralgia and rash," whereas in dengue "myalgia, raised aspartate transaminase and leukopenia" [177]. Some biological factors associated with patients suffering from CHIKV are lymphopenia, increased level of function of liver enzymes, anemia and elevated creatinine level [175, 176].

4.7.1.3.3 Treatment

Commonly, acetoaminophen or nonsteroidal anti-inflammatory drugs are prescribed for the treatment of chikungunya fever. Efficacy of Chloroquine in patient suffering from acute chikungunya fever has been estimated between 300–600 mg.

4.7.2 Sandfly-borne diseases

The causative agent for leishmaniasis is *Leishmania infantum*, infection in humans occurs when female sandfly bites. Biting activity, diapause and maturation functions depend on the temperature [180]. Sandflies are distributed in southern Europe at latitude of 45°N, but now also found in latitude of 49°N. [179]. From ancient times, vector sandflies are diversified from Mediterranean toward northward during the period of postglacial phase; on the morphological basis, samples were also found in "France and northeast Spain," and currently sandflies were spotted in "northern Germany" [180]. The rate of biting in "European sandflies" depends on the seasons, and in most of the regions biting activity is most prevalent in summer months. Currently, sandfly vectors are having significant level in comparison to *L. infantum*, and infection is mostly found in dogs in central and northern Europe. Thus, multifaceted weather and ecological uses (land use) shift the spreading of leishmaniasis in Europe [181].

4.7.2.1 Epidemiology

Phleboviruses are having RNA virus and also three types of genomes (S, M and L segments), three nucleocapsids that are produced within cytosol of a cell. It includes 60 virus which contains different antigens in the serum, which are mainly categorized into groups, viz. the "sandfly fever and the Uukuniemi" [184]. "Sandfly fever viruses" include the "Naples serocomplex" and the "Sicilian serocomplex," which are the two main serocomplexes associated with human diseases. "Karimabad virus," "Arabia virus," "Massilia virus," "Punique virus," "Tehran virus," "Toscana virus (TOSV)" and "unclassified sandfly fever Naples virus" are among the "sandfly fever Naples virus (SFNV) serogroup" by using proteomic assessment. Even though Arabia virus is categorized within the Naples serocomplex [185]. Sandfly fever Sicilian virus (SFSV) is a well-identified serotype having a wide range of geographical distribution. Sandfly Cyprus virus (SFCV) and sandfly Turkey viruses (SFTV) are considered as an alternative of "SFSV; the SFTV," which reveals 98% and 91.8% nucleotide similarities in S segment with SFCV and SFSV prototype strain Sabin, respectively [184]. Recently, Granada virus has been spotted in Spain and has been known that it is close with the SFNV group [185]. The infection is transmitted when phlebotomine sandflies bite the human beings [1988]. The key vector for "SFSV" is *Phlebotomus papatasi*, *P. perfiliewi* is a vector for TOSV and *P. major sensu* is a vector for SFTV [185], and infection mostly spreads in the summer season [186].

4.7.2.2 Diagnosis

Common and predictable techniques for the examination of microscopic examination and culture maintenance is a typical method for examination of leishmaniasis; however, there are some highly sophisticated techniques for the diagnosis which includes molecular assessment [187]. Serological diagnostic method is used for recognition of antibodies within the circulatory blood serum. Immunochromatographic dipstick tests and freeze-dried antigen are used for the detection of visceral leishmaniasis.

4.7.2.2.1 Treatment

In different countries, including India, South America and the Mediterranean, single dose (success rate is 95%) of "LAMB" is generally prescribed for curing of visceral leishmaniasis. Indians suffering from visceral leishmaniasis are resistant to pentavalent antimonials. A combined dose of both pentavalent antimonials (available in two formulations: "sodium stibogluconate" and "meglumine antimoniate") and paromomycin is generally prescribed in Africa. Unfortunately, all these drugs are having considerable disadvantages. Oral administration of miltefosine has been found to be the most efficient remedy to cure "visceral leishmaniasis" and "cutaneous leishmaniasis." However, "miltefosine" is associated with slight, viz. gestational effects if prescribed during 3 months of pregnancy. Unfortunately, "miltefosine" does not work during *L. major* or *L. braziliensis*. In 2006, oral dose of miltefosine was given as a standard drug in India followed by "paromomycin" as an injection administered in different low regimens along with single LAMB dose. In East Africa, a combined dose of sodium stibogluconate (20 mg/kg/day) and intramuscular paromomycin (15 mg/kg) for 17 days is prescribed. However, some other remedies are available in some specific cases such as "cutaneous leishmaniasis." However, remedies vary according to the strain, viz. "topical paromomycin" is efficient for "*L. major*", "*L. tropica*", "*L. Mexicana*", "*L. panamensis*" and "*L. braziliensis*" [188]. Pentamidine is efficient for "*L. guyanensis* [191]. Oral fluconazole or itraconazole is effective in "*L. major*" and "*L. tropica*."

4.7.3 Tick-borne diseases

Tick-borne diseases are now becoming a main human well-being concern in highly affected regions of the world. It has become a primary duty of medical practitioners to be aware about the medical symptom of "tick-borne diseases." Because increased frequency of tick-borne diseases increases day by day, large population of the world requires treatment in the form of high-cost medicines. This places a great financial burden on the healthcare system. The life cycle of ticks completes in 2 years, which is divided into three phases: "larva," "nymph" and "adult." Ticks cause various diseases such as Lyme disease, transmitted by nymphal ticks, displaying higher activity in the end of "spring and initial summer" in moderate regions [189].

4.7.3.1 Lyme disease

4.7.3.1.1 Symptoms
Fever, chills, headache, myalgias, joint pain and swollen lymph nodes are common symptoms. In "erythema migrans," "pimples" might emerge when ticks bite. Other symptoms are "severe headaches," "neck stiffness," "visual changes," "mental cloudiness," "fatigue," "arthritis" with "rigorous pain" and swelling (knees and other large joints). "Lyme arthritis," "myalgia," "bone and/or tendon pain," "heart palpitations (Lyme carditis)," "facial palsy," nerve pain, neuropathy, shortness of breath and malaise [190]. In addition, "EM itchiness" might appear on other parts of the body. Headache, neck pain, fever and meningeal signs or meningitis (inflammation of meninges of the brain and spinal cord) symptoms can include changes in mental status, dizziness and short-term memory impairment [190].

4.7.3.2 Diagnostic testing

Diagnostic tests including detection of "IgM or IgG antibodies" in the circulatory blood serum by using a two-tier protocol EIA or IFA are executed initially; further confirmation is done by a Western blot. Usually, a two-tier test for antibodies comes negative in early Lyme disease. In ambiguous cases, Lyme meningitis testing for intrathecal (requiring a spinal tap) IgM or IgG is optional [190].

4.7.3.3 Treatment

Doxycycline (100 mg two times in a day through oral route) is given for 14- to 21-day duration. Medication can be done for 28 days based on the symptoms. Doxycycline shall be used mixed with food to avoid GI irritation. Amoxicillin is prescribed for adults if allergic symptoms appear by using doxycycline. In children, dose of amoxicillin is given for 21 days. Proper course of antibiotic is essential in order to destroy the infection at root level. Spinal fluid is used to examine the neurological Lyme disease and ceftriaxone through intravenous route is given for a long time [190].

4.7.4 Babesiosis

Babesiosis is commonly reported from "North Eastern and upper Mid-Western US." Babesia microti infects red blood cells. About 1,700 cases of babesiosis are reported from 2001 to 2014. It may be also transmitted through blood transfusion during blood donation. Threatened factors are "asplenia," impaired immunity as well as higher age. In critical cases, noticeable symptoms are "thrombocytopenia," "disseminated intravascular coagulation," "end organ failure" and fatality [193]. Signs and symptoms of Lyme disease are fever, chills, diaphoresis, fatigue, malaise and joint pain. GI symptoms can include nausea and anorexia. Splenomegaly and hepatomegaly can also occur [190].

4.7.4.1 Diagnostic testing

Lyme disease is detected by "intraerythrocytic babesia parasites" observed under "light microscope" by preparing a thin blood film, classically "Maltese cross" on red blood cells. Additional investigative assessment including PCR examination for "Babesia microti" and indirect fluorescent antibody (IFA) for Babesia-specific antibody titer could be consistently done. Common detection tests are hematocrit, thrombocytopenia and elevated liver transaminase values [190].

4.7.4.2 Treatment

Combined oral dose of "atovaquone" (at a dose of 750 mg for 12 h) and "azithromycin" (500–1,000 mg/day for 7–10 days) is used. Further, combined "clindamycin" through intravenous route and "oral quinine" might be prescribed for 7–10 days in patients suffering from serious symptoms. Sometimes, "exchange transfusions" might be important under critical conditions [190].

4.7.5 Anaplasmosis

Anaplasmosis (*Anaplasma phagocytophilum*) is transferred through "Ixodes Scapularis tick" in the "northeastern" and "higher Midwest" areas of the United States in geographic areas having the incidence of Lyme disease. In 2010, 1,700 of anaplasmosis are reported [190]. The latent period for anaplasmosis ranges between 1 and 2 weeks. Clinical signs and symptoms are like other tick-borne diseases such as anaplasmosis, viz. chills, GI symptoms and rarely itchiness [190].

4.7.5.1 Diagnostic testing

Antibodies against "*A. phagocytophilum*" are generally detected within 7–10 days following the onset of infection. Affirmative test shows four times increase in "IgG-specific antibody titer by immunofluorescence assay" in two circulatory blood serum samples (second sample taken 2–4 weeks). Other examination parameters include leukopenia, increased liver-specific biomarkers and anemia. Nevertheless, presence of "morulae" in the cytosol of granulocytes could be seen [190].

4.7.5.2 Treatment

Treatment for anaplasmosis includes "doxycycline" (100 mg/mg 2 times in a day, orally or "intravenously" for 5–7 days). Doxycycline is also prescribed for children (2.2 mg/kg 2 times in a day).

4.7.6 Southern tick-associated rash illness (STARI)

"Lone star tick (*Amblyomma americanum*)" is a main vector of tick, which causes these cuts that are comparatively smaller than "Ixodes tick species-I," "*I. scapularis*" or "*I. pacificus*." These are main vectors for the transmission of "Lyme disease" in the United States. It is still confirmed that whether *Borrelia burgdorferi* causes rashes. This medical creature differs from "Lyme disease" and is well known as "Southern tick-associated rash illness" or "STARI" [194]. Characteristics of STARI are the Lyme-like lesion. Commonly, rashes become visible within 7 days after the biting by "lone star tick," and "Lyme lesion" spreads in the form of circle or oval shape. The symptoms of STARI include "fever," "headache," "stiff neck" and myalgia [191].

4.7.6.1 Diagnostic testing

B. lonestari is usually examined by PCR when "lone star ticks" is detached from a person, and in lone star ticks as reported in epidemiologic findings [191].

4.7.6.2 Treatment

Ticks should be amputed from adults and children as soon as detected, and the same procedure should be applied for other domestic animals including cats, dogs and horses [192].

References

[1]   McKerrow JH, Caffrey C, Kelly B, Loke PN, Sajid M, Proteases in parasitic diseases, Annu Rev Pathol Mech Dis, 2006, 1, 497–536.
[2]   Burrell CJ, Howard CR, Murphy FA, Chapter 7 – Pathogenesis of virus infections, Burrell CJ, Howard CR, Murphy FA (eds.), Fenner and White's Medical Virology, Fifth Edition, Academic Press, Elsevier, USA, 2017, 77–104.
[3]   Andrews KT, Fisher G, Skinner-Adams TS, Drug repurposing and human parasitic protozoan diseases, Int J Parasitol: Drugs Drug Resist, 2014, 4(2), 95–111.
[4]   Lustigman S, Prichard RK, Gazzinelli A, Grant WN, Boatin BA, McCarthy JS, Basanez MG, A research agenda for helminth diseases of humans: the problem of helminthiases, PLoS Negl Trop Dis, 2012, 6(4), e1582.
[5]   Ryan KJ, Chapter 49 – Pathogenesis and diagnosis of parasitic infection, Sherris Medical Microbiology, 7e, 2018, 370.
[6]   Wang Y, Pathogenesis of parasitic diseases, Li H (eds.), Radiology of Parasitic Diseases, Dordrecht, Springer, 2017, 9–11.
[7]   Brener Z, Gazzinelli RT, Immunological control of Trypanosoma cruzi infection and pathogenesis of Chagas' disease, Int Arch Allergy Immunol, 1997, 114(2), 103–10.
[8]   Bonney KM, Luthringer DJ, Kim SA, Garg NJ, Engman DM, Pathology and pathogenesis of Chagas heart disease, Annu Rev Pathol: Pathol Mech Dis, 2019, 14, 421–47.
[9]   A R Jr, Rassi A, de Rezende JM, American trypanosomiasis (Chagas disease), Infect Dis Clin North Am, 2012, 26, 275–91.
[10]   Lopez-Munoz RA, Molina-Berríos A, Campos-Estrada C, Abarca-Sanhueza P, Urrutia-Llancaqueo L, Pena-Espinoza M, Maya JD, Inflammatory and pro-resolving lipids in trypanosomatid infections: a key to understanding parasite control, Front Microbiol, 2018, 9. 1961.

[11]   Seed JR, Chapter 78 – Protozoa: pathogenesis and defenses, Baron S (eds.), Medical
       Microbiology, 4th edn., Galveston, University of Texas Medical Branch at Galveston, 1996.
[12]   Pina-Vazquez C, Reyes-Lopez M, Ortíz-Estrada G, De La Garza M, Serrano-Luna J, Host-parasite
       interaction: parasite-derived and-induced proteases that degrade human extracellular matrix,
       J Parasitol Res, 2012, 2012, 1–25.
[13]   Ndao M, Diagnosis of parasitic diseases: old and new approaches, Interdiscip Perspect Infect
       Dis, 2009, 2009, 1–16.
[14]   Rosenblatt JE, Reller LB, Weinstein MP, Laboratory diagnosis of infections due to blood and
       tissue parasites, Clin Infect Dis, 2009, 49(7), 1103–08.
[15]   Foreyt WJ, Diagnostic parasitology, Vet Clin North Am SAP, 1989, 19(5), 979–1000.
[16]   Hancock K, Tsang VC, Development and optimization of the FAST-ELISA for detecting
       antibodies to Schistosoma mansoni, J Immunol Methods, 1986, 92(2), 167–76.
[17]   Pappas MG, Hajkowski R, Hockmeyer WT, Dot enzyme-linked immunosorbent assay
       (Dot-ELISA): A micro technique for the rapid diagnosis of visceral leishmaniasis, J Immunol
       Methods, 1983, 64(1–2), 205–14.
[18]   Shokoples SE, Ndao M, Kowalewska-Grochowska K, Yanow SK, Multiplexed real-time PCR
       assay for discrimination of Plasmodium species with improved sensitivity for mixed
       infections, J Clin Microbiol, 2009, 47(4), 975–80.
[19]   Burbelo PD, Goldman R, Mattson TL, A simplified immunoprecipitation method for quantitatively
       measuring antibody responses in clinical sera samples by using mammalian-produced Renilla
       luciferase-antigen fusion proteins, BMC Biotechnol, 2005, 5, article 22, 1–10.
[20]   Andrews RH, Chilton NB, Multilocus enzyme electrophoresis: A valuable technique for
       providing answers to problems in parasite systematic, Int J Parasitol, 1999, 29(2), 213–53.
[21]   Muldrew KL, Molecular diagnostics of infectious diseases, Curr Opin Pediatr, 2009, 21(1), 102–11.
[22]   Parida MM, Sannarangaiah SP, Dash KP, Rao VL, Morita K, Loop mediated isothermal
       amplification (LAMP): a new generation of innovative gene amplification technique;
       perspectives in clinical diagnosis of infectious diseases, Rev Med Virol, 2008, 18(6), 407–21.
[23]   Tait BD, Hudson F, Cantwell L, et al., Review article: Luminex technology for HLA antibody
       detection in organ transplantation, Nephrology, 2009, 14(2), 247–54.
[24]   De Waal T, Advances in diagnosis of protozoan diseases, Vet Parasitol, 2012, 189(1), 65–74.
[25]   Garcia LS, Practical Guide to Diagnostic Parasitology, Washington, American Society for
       Microbiology, 1999, 349 pp.
[26]   Cheesbrough M, Direct Laboratory Practice in Tropical Countries (Part-1), New York,
       Cambridge University Press, 2009, pp., 29–35.
[27]   Khurana S, Sethi S, Laboratory diagnosis of soil transmitted helminthiasis, Trop Parasitol,
       2017, 7(2), 86.
[28]   Requena-Mendez A, Chiodini P, Bisoffi Z, Buonfrate D, Gotuzzo E, Muñoz J, The laboratory
       diagnosis and follow up of strongyloidiasis: a systematic review, PLoS Negl Trop Dis, 2013,
       7, e2002.
[29]   Carter WJ, Yan Z, Cassai ND, Sidhu GS, Detection of extracellular forms of babesia in the
       blood by electron microscopy: a diagnostic method for differentiation from *Plasmodium
       falciparum*, Ultrastruct Pathol, 2003, 27(4), 211–16.
[30]   Diez M, Favaloro L, Bertolotti A, Burgos JM, Vigliano C, Lastra MP, Levin MJ, Arnedo A, Nagel
       C, Schijman AG, Favaloro RR, Usefulness of PCR strategies for early diagnosis of Chagas'
       disease reactivation and treatment follow-up in heart transplantation, Am J Transplant, 2007,
       7(6), 1633–40.
[31]   Hancock K, Tsang VC, Development and optimization of the FAST-ELISA for detecting
       antibodies to Schistosoma mansoni, J Immunol Methods, 1986, 92(2), 167–76.

[32] Campbell GH, Aley SB, Ballou WR, Hall T, Hockmeyer WT, Hoffman SL, Hollingdale MR, Howard RJ, Lyon JA, Nardin EH, Nussenzweig RS, Nussenzweig V, Tsang VCW, Weber JL, Wellems TE, Young JF, Zavala F, Use of synthetic and recombinant peptides in the study of host-parasite interactions in the malarias, Am J Trop Med Hyg Title, 1987, 37(3), 428–44.

[33] Hillyer GV, de Galanes MS, Rodriguez-Perez J, et al., Use of the Falcon^{TM} assay screening test-enzyme-linked immunosorbent assay (FAST-ELISA) and the enzyme linked immunoelectrotransfer blot (EITB) to determine the prevalence of human fascioliasis in the Bolivian Altiplano, Am J Trop Med Hyg, 1992, 46(5), 603–09.

[34] Maddison SE, The present status of serodiagnosis and seroepidemiology of schistosomiasis, Diagn Microbiol Infect Dis, 1987, 7(2), 93–105.

[35] Ko RC, Ng TF, Evaluation of excretory/secretory products of larval Taenia solium as diagnostic antigens for porcine and human cysticercosis, J Helminthol, 1998, 72(2), 147–54.

[36] Pappas MG, Recent applications of the Dot-ELISA in immunoparasitology, Vet Parasitol, 1988, 29(2–3), 105–29.

[37] World Health Organization, The Use of Malaria Rapid Diagnostic Tests, Manila, WHO Regional Office for the Western Pacific, 2006.

[38] Murray CK, Gasser RA Jr, Magill AJ, Miller RS, Update on rapid diagnostic testing for malaria, Clin Microbiol Rev, 21(1), 97–110, 2008.

[39] Drakeley C, Reyburn H, Out with the old, in with the new: the utility of rapid diagnostic tests for malaria diagnosis in Africa, Trans R Soc Trop Med Hyg, 2009, 103(4), 333–37.

[40] Bell D, Peeling RW, Evaluation of rapid diagnostic tests: malaria, Nat Rev Microbiol, 2006, 4(9supp), S34.

[41] Agatsuma T, Terasaki K, Yang L, Blair D, Genetic variation in the triploids of Japanese Fasciola species, and relationships with other species in the genus, J Helminthol, 1994, 68(3), 181–86.

[42] Lymbery AJ, Thompson RCA, Electrophoretic analysis of genetic variation in Echinococcus granulosus from domestic hosts in Australia, Int J Parasitol, 1988, 18(6), 803–11.

[43] Barnabe C, Yaeger R, Pung O, Tibayrenc M, Trypanosoma cruzi: a considerable phylogenetic divergence indicates that the agent of Chagas disease is indigenous to the native fauna of the United States, Exp Parasitol, 2001, 99(2), 73–79.

[44] Murphy RW, Sites JW, Buth DG, Haufler CH, Proteins I: isoenzyme electrophoresis, Molecular Systematics, Hillis DM, Moritz C (eds.), Sunderland, Mass., Sinauer Associates Inc. Publishers, 1990, 45–126.

[45] Saiki RK, Gelfand DH, Stoffel S, et al., Primer-directed enzymatic amplification of DNA with a thermostable DNA polymerase, Science, 1988, 239, 487–91.

[46] Mens P, Spieker N, Omar S, Heijnen M, Schallig H, Kager PA, Is molecular biology the best alternative for diagnosis of malaria to microscopy? A comparison between microscopy, antigen detection and molecular tests in rural Kenya and urban Tanzania, Trop Med Int Health, 2007, 12(2), 238–44.

[47] Maddocks S, Jenkins R, Understanding PCR: a Practical Bench-top Guide, Academic Press, USA, 2016.

[48] Jia Y, Real-time PCR, Methods in Cell Biology, 112, Academic Press, USA, 2012, 55–68.

[49] van Lieshout L, Verweij JJ, Newer diagnostic approaches to intestinal protozoa, Curr Opin Infect Dis, 2010, 23(5), 488–93.

[50] Gasser RB, Molecular tools – advances, opportunities and prospects, Vet Parasitol, 2006, 136(2), 69–89.

[51] Farcas GA, Soeller R, Zhong K, Zahirieh A, Kain KC, Real-time polymerase chain reaction assay for the rapid detection and characterization of chloroquine-resistant Plasmodium falciparum malaria in returned travelers, Clin Infect Dis, 2006, 42(5), 622–27.

[52] Nazeer JT, Khalifa KES, Von Thien H, El-Sibaei MM, Abdel-Hamid MY, Tawfik RAS, Tannich E, Use of multiplex real-time PCR for detection of common diarrhea causing protozoan parasites in Egypt, Parasitol Res, 2013, 112(2), 595–601.

[53] Njiru ZK, Mikosza ASJ, Matovu E, et al., African trypanosomiasis: Sensitive and rapid detection of the subgenus Trypanozoon by loop-mediated isothermal amplification (LAMP) of parasite DNA, Int J Parasitol, 2008, 38(5), 589–99.

[54] Lucchi NW, Demas A, Narayanan J, Sumari D, Kabanywanyi A, Kachur SP, Udhayakumar V, Real-time fluorescence loop mediated isothermal amplification for the diagnosis of malaria, PloS One, 2010, 5(10), e13733.

[55] Kuboki N, Inoue N, Sakurai T, Di Cello F, Grab DJ, Suzuki H, Sugimoto C, Igarashi I, Loop-mediated isothermal amplification for detection of African trypanosomes, J Clin Microbiol, 2003, 41, 5517–24.

[56] Thekisoe OM, Kuboki N, Nambota A, Fujisaki K, Sugimoto C, Igarashi I, Yasuda J, Inoue N, Species-specific loop-mediated isothermal amplification (LAMP) for diagnosis of trypanosomosis, Acta Trop, 2007, 102, 182–89.

[57] Han ET, Watanabe R, Sattabongkot J, et al., Detection of four Plasmodium species by genus- and species-specific loop-mediated isothermal amplification for clinical diagnosis, J Clin Microbiol, 2007, 45(8), 2521–28.

[58] Alhassan A, Thekisoe OMM, Yokoyama N, et al., Development of loop-mediated isothermal amplification (LAMP) method for diagnosis of equine piroplasmosis, Vet Parasitol, 2007, 143(2), 155–60.

[59] Nkouawa A, Sako Y, Nakao M, Nakaya K, Ito A, Loop-mediated isothermal amplification method for differentiation and rapid detection of Taenia species, J Clin Microbiol, 2009, 47(1), 168–74.

[60] Cowan LS, Diem L, Brake MC, Crawford JT, Transfer of a Mycobacterium tuberculosis genotyping method, spoligotyping, from a reverse line-blot hybridization, membrane-based assay to the Luminex multianalyte profiling system, J Clin Microbiol, 2004, 42(1), 474–77.

[61] Li W, Zhang N, Gong P, Cao L, Li J, Su L, et al., A novel multiplex PCR coupled with Luminex assay for the simultaneous detection of Cryptosporidium spp., Cryptosporidium parvum and Giardia duodenalis, Vet Parasitol, 2010, 173(1–2), 11–18.

[62] McNamara DT, Kasehagen LJ, Grimberg BT, Cole-Tobian J, Collins WE, Zimmerman PA, Diagnosing infection levels of four human malaria parasite species by a polymerase chain reaction/ligase detection reaction fluorescent microsphere-based assay, Am J Trop Med Hyg, 2006, 74(3), 413–21.

[63] Mills AE, Goldsmid JM, Intestinal protozoa, Tropical Pathology, Berlin, Heidelberg, Springer 1995, 477–556. Mills and Goldsmith are editors are book.

[64] Bogitsh BJ, Carter CE, Oeltmann TN, Human Parasitology, Academic Press, Elsevier, USA, 2019.

[65] Kissinger P, Trichomonas vaginalis: a review of epidemiologic, clinical and treatment issues, BMC Infect Dis, 2015, 15, 307.

[66] Haque R, Huston CD, Hughes M, Houpt E, Petri WA Jr, Amebiasis, N Engl J Med, 2003, 348, 1565–73.

[67] Houpt E, Hung CC, Petri W, Entamoeba histolytica (amebiasis). Hunter's Tropical Medicine and Emerging Infectious Diseases, 9th edn., Magill AJ, Ryan ET, Solomon T, Hill DR, (eds.), Philadelphia, Saunders, 2012, 659.

[68] Chalmers RM, Chapter eighteen – Entamoeba histolytica, (eds), Percival SL, Yates MV, Williams DW, Chalmers RM, Gray NF, Microbiology of Waterborne Diseases, 2nd edn., Academic Press, London, UK, 2014, 355–373.

[69] Hamano S, Petri WA Jr, Kris H, Protozoan diseases: amebiasis, International Encyclopedia of Public Health, Oxford, Academic, Press, United Kingdoms, 2008, 335–41.

[70] Hanevik K, Dizdar V, Langeland N, Hausken T, Development of functional gastrointestinal disorders after *Giardia lamblia* infection, BMC Gastroenterol, 2009, 9(1), 27.

[71] Adam RD, Biology of Giardia lamblia, Clin Microbiol Rev, 2001, 14, 447–75.

[72] Carranza PG, Lujan HD, New insights regarding the biology of Giardia lamblia, Microbes Infect, 2010, 12, 71–80.

[73] Ankarklev J, Jerlstrom-Hultqvist J, Ringqvist E, Troell K, Svard SG, Behind the smile: cell biology and disease mechanisms of Giardia species, Nat Rev Microbiol, 2010, 8, 413–22.

[74] Lopez-Romero G, Quintero J, Astiazarán-García H, Velazquez C, Host defences against Giardia lamblia, Parasite Immunol, 2015, 37, 394–406.

[75] Lujan HD, Giardia SS, A Model Organism, Austria, Springer Wein New York, 2011, 1–417.

[76] Percival SL, Chalmers R, Embrey M, Hunter PR, Sellwood J, Wyn-Jones P, Microbiology of Waterborne Diseases Cryptosporodium, eUK UK: Elsevier academic press, 2004.

[77] Chalmers RM, Davies AP, Mini-review: clinical cryptosporidiosis, Exp Parasitol, 2010, 124, 138–46.

[78] Chen XM, LaRusso NF, Cryptosporidiosis and the pathogenesis of AIDS cholangiopathy, Semin Liver Dis, 2002, 22, 277–89.

[79] Fayer R, General biology, Fayer R, Xiao L, (eds.), Cryptosporidium and Cryptosporidiosis, Boca Raton, CRC Press, 2008, pp., 1–42.

[80] Jenkins MB, Eaglesham BS, Anthony LC, et al., 2010, Significance of wall structure, macromolecular composition, and surface polymers to the survival and transport of Cryptosporidium parvum oocysts, Appl Environ Microbiol, 76, 1926–34.

[81] Current WL, Reese NC, A comparison of endogenous development of three isolates of Cryptosporidium in suckling mice, J Protozool, 1986, 33, 98–108.

[82] Chalmers RM, Chapter sixteen – cryptosporidium, Percival SL, Yates MV, Williams DW, Chalmers RM, Gray NF (eds.), Microbiology of Waterborne Diseases, 2nd edn., Academic Press, United Kingdom, London, 2014, 287–326.

[83] Chalmers RM, Davies AP, Minireview: clinical cryptosporidiosis, Exp Parasitol, 2010, 124(1) 138–46.

[84] Lunde CS, Stebbins EE, Jumani RS, Hasan MM, Miller P, Barlow J, Freund YR, Berry P, Stefanakis R, Gut J, Rosenthal PJ, Love MS, McNamara CW, Easom E, Plattner JJ, Jacobs RT, Huston CD Identification of a potent benzoxaborole drug candidate for treating cryptosporidiosis, Nat Commun, 2019, 10(1), 1–11.

[85] Page SW, Antiparasitic drugs, Small Anim Clin Pharmacol, 2008, 2, 198–260.

[86] World Malaria Report. 2015. http://www.who.int/malaria/publications/world-malaria-report-2015/report/en/.

[87] Maser P, Wittlin S, Rottmann M, Wenzler T, Kaiser M, Brun R, Antiparasitic agents: new drugs on the horizon, Curr Opin Pharmacol, 2012, 12(5), 562–66.

[88] Martínez-Palomo A, Espinosa-Cantellano M, Tsutsumi GR, Tsutsumi V, Chapter 13 – Amebic liver abscess, Muriel P (eds.), Liver Pathophysiology, Academic Press, United Kingdom, 2017, Pages 181–186.

[89] Campbell S, Soman-Faulkner K, Antiparasitic Drugs, StatPearls, Treasure Island (FL), StatPearls Publishing, 2019.

[90] Cudmore SL, Garber GE, Prevention or treatment: the benefits of Trichomonas vaginalis vaccine, J Infect Public Health, 2010, 3(2), 47–53.

[91] Rais S, Perianin A, Lenoir M, Sadak A, Rivollet D, Paul M, Deniau M, Sodium stibogluconate (Pentostam) potentiates oxidant production in murine visceral leishmaniasis and in human blood, Antimicrob Agents Chemother, 2000, 44(9), 2406–10.

[92] Yesilova Y, Surucu HA, Ardic N, Aksoy M, Yesilova A, Oghumu S, Satoskar AR, Meglumine antimoniate is more effective than sodium stibogluconate in the treatment of cutaneous leishmaniasis, J Dermatolog Treat, 2016, 27(1), 83–87.

[93] Aflatoonian MR, Sharifi I, Aflatoonian B, Bamorovat M, Heshmatkhah A, Babaei Z, Ghasemi Nejad Almani P, Mohammadi MA, Salarkia E, Aghaei Afshar A, Sharifi H, Sharifi F, Khosravi A, Khatami M, Arefinia N, Fekri A, Farajzadeh S, Khamesipour A, Mohebali M, Gouya MM, Shirzadi MR, Varma RS, Associated-risk determinants for anthroponotic cutaneous leishmaniasis treated with meglumine antimoniate: a cohort study in Iran, PLoS Negl Trop Dis, 2019, Jun, 13(6), e0007423.

[94] Rios L, Campos EE, Menon R, Zago MP, Garg NJ, Epidemiology and pathogenesis of maternal-fetal transmission of Trypanosoma cruzi and a case for vaccine development against congenital Chagas disease, Biochim Biophys Acta (BBA)-Mol Basis Dis, 2019, 1866, 165591.

[95] Thomas JA, Baker N, Hutchinson S, Dominicus C, Trenaman A, Glover L, Alsford S, Horn D, Insights into antitrypanosomal drug mode-of-action from cytology-based profiling, PLoS Negl Trop Dis, 2018 Nov, 12(11), e0006980.

[96] Balfour JA, McClellan K, Topical eflornithine, Am J Clin Dermatol, 2001, 2(3), 197–201; discussion 202.

[97] Morgan HP, McNae IW, Nowicki MW, Zhong W, Michels PA, Auld DS, Fothergill-Gilmore LA, Walkinshaw MD, The trypanocidal drug suramin and other trypan blue mimetics are inhibitors of pyruvate kinases and bind to the adenosine site, J Biol Chem, 2011, Sep 09, 286(36), 31232–40.

[98] Fairlamb AH, Horn D, Melarsoprol resistance in African trypanosomiasis, Trends Parasitol, 2018, Jun, 34(6) 481–92.

[99] Cupit PM, Cunningham C, What is the mechanism of action of praziquantel and how might resistance strike?, Future Med Chem, 2015, 7(6), 701–05.

[100] Chen W, Mook RA, Premont RT, Wang J, Niclosamide: beyond an antihelminthic drug, Cell Signal, 2018, Jan, 41, 89–96.

[101] Srivastava IK, Rottenberg H, Vaidya AB, Atovaquone, a broadspectrum antiparasitic drug, collapses mitochondrial membrane potential in a malarial parasite, J Biol Chem, 1997, 272, 3961–66.

[102] Jacobs RT, Antiparasitic agents, Comp Med Chemi, III, 2017, 717–50.

[103] Ghosh M, Roy K, Roy S, Immunomodulatory effects of antileishmanial drugs, J Antimicrob Chemother, 2013, 68(12), 2834–38.

[104] Bhargava P, Singh R, Developments in diagnosis and antileishmanial drugs, Interdiscip Perspect Infect Dis, 2012, 2020, 1–13.

[105] Croft SL, Chatelain E, Barrett MP, Antileishmanial and antitrypanosomal drug identification, Emerg Top Life Sci, 2017, 1, 613–20.

[106] Laing R, Gillan V, Devaney E, Ivermectin – old drug, new tricks?, Trends Parasitol, 2017, 33(6), 463–72.

[107] Brun R, Blum J, Chappuis F, Burri C, Human African trypanosomiasis, Lancet, 2010, 375 (9709), 148–59.

[108] Matthews KR, The developmental cell biology of Trypanosoma brucei, J Cell Sci, 2005, 118, 283–90.

[109] Hemmige V, Tanowitz H, Sethi A, Trypanosoma cruzi infection: a review with emphasis on cutaneous manifestations, Int J Dermatol, 2012, 51(5), 501–08.

[110] Simarro PP, Franco J, Diarra A, Postigo JA, Jannin J, Update on field use of the available drugs for the chemotherapy of human African trypanosomiasis, Parasitology, 2012, 139(7), 842–46.

[111] Pal M, Zoonoses, 1997, New Delhi, India, RM publisher and distributor.

[112] Assimina Z, Charilaos K, Fotoula B, Leishmaniasis: an overlooked public health concern, Health Sci J, 2008, 2, 196–205.

[113] Dantas-Torres F, The role of dogs as reservoirs of Leishmania parasites, with emphasis on Leishmania (Leishmania) infantum and Leishmania (Viannia) braziliensis, Vet Parasitol, 2007, 149, 139–46.

[114] Center for Food Security and Public Health (CFSPH), Leishmaniasis (Cutaneous and Visceral), 2009, Lowa, Lowa State of University, College of Veterinary Medicine.

[115] Koutis CH, Special Epidemiology, Editions, 2007, Athens, Greece, Technological Educational Institute of Athens.

[116] Bañuls A, Hide M, Prugnolle F, *Leishmania* and the leishmaniases:A parasite genetic updates and advances in taxonomy, epidemiology and pathogenicity in humans, Adv Parasitol, 2007, 64, 1–109.

[117] Getachew T, Tadesse A, Yoseph M, Zenebe A, Abere B, et al. Internal medicine lecture notes for health officers, Ethiopian in collaboration with the Ethiopia Public Health Training Initiative. The Carter Center, the Ethiopia Ministry of Health, and the Ethiopia Ministry of Education, 2006, 56–63.

[118] Hide M, Bucheton B, Kamhawi S, Bras-Gonçalves R, Sundar S, et al., Understanding Human Leishmaniasis: The Need for an Integrated Approach in Encyclopedia of Infectious Diseases Book of Microbiology, John Wiley and Sons Inc, 2007, 87–107, United States, Hoboken, New Jersey.

[119] Chappuis F, Sundar S, Hailu A, Ghalib H, Rijal S, et al., Visceral leishmaniasis: What are the needs for diagnosis treatment and control?, Nat Rev Microbiol, 2007, 5, 873–82.

[120] Herwaldt BL, Harrison's Principles of Internal Medicine, 16th edn, leishmaniasis, 2005, 1233–38, New York, USA: Harrison House Publishers.

[121] Howes RE, Battle KE, Mendis KN, Smith DL, Cibulskis RE, Baird JK, Hay SI, Global Epidemiology of *Plasmodium vivax*, Am J Trop Med Hyg, 2016, 95, 15–34.

[122] Dayananda KK, Achur RN, Gowda DC, Epidemiology, drug resistance, and pathophysiology of *Plasmodium vivax* malaria, J Vector Borne Dis, 2018, 55(1), 1–8.

[123] Hemmer CJ, Holst FG, Kern P, Chiwakata CB, Dietrich M, Reisinger EC, Stronger host response per parasitized erythrocyte in *Plasmodium vivax* or *ovale* than in *Plasmodium falciparum* malaria, Trop Med Int Health, 2006, 11(6), 817–23.

[124] Dondorp A, Nosten F, Stepniewska K, Day N, White N, South East Asian Quinine Artesunate Malaria Trial (SEAQUAMAT) group. Artesunate versus quinine for treatment of severe falciparum malaria: A randomised trial, Lancet, 2005, 366(9487), 717–25.

[125] Jones KL, Donegan S, Lalloo DG, Artesunate versus quinine for treating severe malaria., Cochrane Database Syst Rev, 2007, 17(4), CD005967.

[126] Hess KM, Goad JA, Arguin PM, Intravenous artesunate for the treatment of severe malaria, Ann Pharmacother, 2010, 44(7–8), 1250–58.

[127] Garcia LS, Practical Guide to Diagnostic Parasitology, American Society for Microbiology Press, 2009, John Wiley & Sons, ASM Press, Washington, DC, USA.

[128] Thompson DP, Geary TG, Helminth surfaces: Structural, molecular and functional properties, Molecular Medical Parasitology, 2003, 297–338, Academic Press, ELSEVIER, USA.

[129] Castro GA, Helminths: Structure, Classification, Growth, and Development, Medical Microbiology, 4th ed, 2011, Galveston, University of Texas Medical Branch at Galveston, 1996.

[130] Jeffrey HC, Leach RM: Atlas of Medical Helminthology and Protozoology., Edinburgh, Churchill Livingstone, 1968.

[131] Hagel I, Giusti T, Ascaris lumbricoides: An overview of therapeutic targets, Infect Disord-Drug Targets (Formerly Curr Drug Targets Infect Disord), 2010, 10(5), 349–67.

[132] Gottstein B, Pozio E, Nöckler K, Epidemiology, diagnosis, treatment, and control of trichinellosis, Clin Microbiol Rev, 2009, 22(1), 127–45.

[133] Murrell KD, Pozio E, Trichinellosis: The zoonosis that won't go quietly, Int J Parasitol, 2000, 30(12–13), 1339–49.

[134] Capó V, Despommier DD, Clinical aspects of infection with Trichinella spp, Clin Microbiol Rev, 1996, 9(1), 47–54.

[135] Gyorkos TW, Larocque R, Casapia M, Gotuzzo E, Lack of risk of adverse birth outcomes after deworming in pregnant women, Pediatr Infect Dis J, 2006, 25(9), 791–194.

[136] Halpenny CM, Paller C, Koski KG, Valdés VE, Scott ME, Regional, household and individual factors that influence soil transmitted helminth reinfection dynamics in preschool children from rural indigenous Panamá, PLoS Negl Trop Dis, 2013, 7(2), p.e2070.

[137] Fürst T, Keiser J, Utzinger J, Global burden of human food-borne trematodiasis: A systematic review and meta-analysis, Lancet Infect Dis, 2012, 12(3), 210–21.

[138] Brooker S, Peshu N, Warn PA, Mosobo M, Guyatt HL, Marsh K, Snow RW, The epidemiology of hookworm infection and its contribution to anaemia among pre-school children on the Kenyan coast, Trans R Soc Trop Med Hyg, 1999, 93(3), 240–46.

[139] Shiferaw MB, Mengistu AD, Helminthiasis: Hookworm infection remains a public health problem in Dera district, South Gondar, Ethiopia, PLoS One, 2015, 10, 12.

[140] Sengchanh K, Manithong V, Peter O, Boungnong B, Soil-transmitted helminth infections and risk factors in preschool children in southern rural Lao People's Democratic Republic, Trans R Soc Trop Med Hyg, 2011, 105(3), 160–66.

[141] Roberts LS, Janovy J, Gerald D, Schmidt & Larry S. Roberts' Foundations of Parasitology, 2009, McGraw-Hill, New York, New York, United States.

[142] Parija SC, Marrie TJ, Koirala S, Trematode Infection, eMedicine J, May, 2003, 3, 2004.

[143] Doughty BL, Schistosomes and Other Trematodes, Baron S, editor, Medical Microbiology, 4th edition, Galveston, University of Texas Medical Branch at Galveston, 1996, Chapter 88, Available from, https://www.ncbi.nlm.nih.gov/books/NBK8037/. PMID: 21413320 Bookshelf ID: NBK8282.

[144] Jime´nez FA, Gardner SL, Arau´jo A, et al., Zoonotic and human parasites of inhabitants of Cueva de los Muertos Chiquitos, Rio Zape Valley, Durango, Mexico, Journal of Parasitology, 2012, 98, 304–09.

[145] Chai JY, Lee SH, Intestinal trematodes of humans in Korea: Metagonimus, heterophyids and echinostomes, Korean J Parasitol, 1990, 28, Suppl, 103–22.

[146] Qian MB, Utzinger J, Keiser J, Zhou XN, Clonorchiasis, Lancet, 2016, 387(10020), 800–10.

[147] Shin EH, Guk SM, Kim HJ, Lee SH, Chai JY, Trends in parasitic diseases in the Republic of Korea, Trends Parasitol, 2008, 24(3), 143–50.

[148] Petney TN, Andrews RH, Saijuntha W, Wenz-Mücke A, Sithithaworn P, The zoonotic, fish-borne liver flukes Clonorchis sinensis, Opisthorchis felineus and Opisthorchis viverrini, Int J Parasitol, 2013, 43(12–13), 1031–46.

[149] WHO. Control of foodborne trematode infections. Report of a WHO Study Group. WHO Technical Report Survey, 1995;849:1–157.

[150] Cabada MM, White AC Jr, New developments in epidemiology, diagnosis, and treatment of fascioliasis, Curr Opin Infect Dis, 2012, 25(5), 518–22.

[151] Vidamaly S, Choumlivong K, Keolouangkhot V, Vannavong N, Kanpittaya J, Strobel M, Paragonimiasis: A common cause of persistent pleural effusion in Lao PDR, Trans R Soc Trop Med Hyg, 2009, 103(10), 1019–23.

[152] Asnis D, Kazakov J, Toronjadze T, Bern C, Garcia HH, McAuliffe I, Bishop H, Lee L, Grossmann R, Garcia MA, Di John D, Neurocysticercosis in the infant of a pregnant mother with a tapeworm, Am J Trop Med Hyg, 2009, 81(3), 449–51.

[153] Vaidya A, Singhal S, Dhall S, Manohar A, Mahajan H, Asymptomatic disseminated cysticercosis, J Clin Diagn Res, 2013, 7(8), 1761–63.

[154] Madigubba S, Vishwanath K, Reddy GB, Vemuganti GK, Changing trends in ocular cysticercosis over two decades: An analysis of 118 surgically excised cysts, Indian J Med Microbiol, 2007, 25(3), 214–19.

[155] Yang YR, Sun T, Li Z, Zhang J, Teng J, Liu X, Liu R, Zhao R, Jones MK, Wang Y, Wen H, Community surveys and risk factor analysis of human alveolar and cystic echinococcosis in Ningxia Hui Autonomous Region, China Bulletin of the World Health Organization, 2006, 84, 714–21.

[156] Otero-Abad B, Torgerson PR, A systematic review of the epidemiology of echinococcosis in domestic and wild animals, PLoS Negl Trop Dis, 2013, 7(6), 2249.

[157] Rana US, Sehgal S, Bhatia R, Bhardwaj M, Hydatidosis in animals and around Delhi, J Commun Dis, 1986, 18(2), 116–19.

[158] Baumann S, Shi R, Liu W, Bao H, Schmidberger J, Kratzer W, Li W, Worldwide literature on epidemiology of human alveolar echinococcosis: A systematic review of research published in the twenty-first century, Infection, 2019, 30, 1–25.

[159] McManus DP, Thompson RC, Molecular epidemiology of cystic echinococcosis, Parasitology, 2003, 127, 37–51.

[160] Eckert J, Deplazes P, Biological, epidemiological, and clinical aspects of echinococcosis, a zoonosis of increasing concern, Clin Microbiol Rev, 2004, 17(1), 107–35.

[161] Rogers DJ, Randolph SE, Climate change and vector-borne diseases, Adv Parasitol, 2006, 62, 345–81.

[162] Ludwig A, Bicout D, Chalvet-Monfray K, Sabatier P, Modelling the aggressiveness of the Culex modestus, possible vector of West Nile fever in Camargue, as a function of meteorological data, Environn Risques Santé, 2005, 4(2), 109–1013.

[163] Busch MP, Kleinman SH, Tobler LH, Kamel HT, Norris PJ, Walsh I, Matud JL, Prince HE, Lanciotti RS, Wright DJ, Linnen JM, Virus and antibody dynamics in acute West Nile virus infection, J Infect Dis, 2008, 198(7), 984–93.

[164] Murray KO, Walker C, Gould E, The virology, epidemiology, and clinical impact of West Nile virus: A decade of advancements in research since its introduction into the Western Hemisphere, Epidemiol Infect, 2011, 139(6), 807–17.

[165] Bai F, Kong KF, Dai J, Qian F, Zhang L, Brown CR, Fikrig E, Montgometry RR, A paradoxical role for neutrophils in the pathogenesis of West Nile virus, Journal of Infectious Diseases, 2010, 202(12), 1804–12.

[166] Zou S, Foster GA, Dodd RY, Petersen LR, Stramer SL, West Nile fever characteristics among viremic persons identified through blood donor screening, J Infect Dis, 2010, 202(9), 1354–61.

[167] Abbate JL, Determinants of Disease Distribution in a Natural Host-pathogen System, University of Virginia, 2012, ProQuest LLC, 789 EASTEISEN PARKWAY, UNITED STATES.

[168] Anderson JF, Main AJ, Importance of vertical and horizontal transmission of West Nile virus by Culex pipiens in the Northeastern United States, J Infect Dis, 2006, 194(11), 1577–79.

[169] Srivastava R, Ramakrishna C, Cantin E, Anti-inflammatory activity of intravenous immunoglobulins protects against West Nile virus encephalitis, J Gen Virol, 2015, 96(Pt 6), 1347.

[170] Beltrame A, Angheben A, Bisoffi Z, Monteiro G, Marocco S, Calleri G, Lipani F, Gobbi F, Canta F, Castelli F, Gulletta M, Imported chikungunya infection, Italy, Emerg Infect Dis, 2007, 13(8), 1264–66.

[171] Medlock JM, Avenell D, Barrass I, Leach S, Analysis of the potential for survival and seasonal activity of Aedes albopictus (Diptera: Culicidae) in the United Kingdom, J Vector Ecol, 2006, 31(2), 292–304.

[172] Knudsen AB, Global distribution and continuing spread of Aedes albopictus, Parassitologia, 1995, 37(2–3), 91–97.

[173] Yadav P, Gokhale MD, Barde PV, Singh DK, Mishra AC, Mourya DT, Experimental transmission of Chikungunya virus by Anopheles stephensi mosquitoes, Acta Virol, 2003, 47(1), 45–47.

[174] Sissoko D, Ezzedine K, Moendandzé A, Giry C, Renault P, Malvy D, Field evaluation of clinical features during chikungunya outbreak in Mayotte, 2005–2006, Trop Med Int Health, 2010, 15(5), 600–07.

[175] Nkoghe D, Kassa RF, Caron M, Grard G, Mombo I, Bikie B, Paupy C, Becquart P, Bisvigou U, Leroy EM, Clinical forms of chikungunya in Gabon, 2010, PLoS Negl Trop Dis, 2012, 6(2), e1517.

[176] Staikowsky F, Talarmin F, Grivard P, Souab A, Schuffenecker I, Le Roux K, Lecuit M, Michault A, Prospective study of Chikungunya virus acute infection in the Island of La Reunion during the 2005–2006 outbreak, PloS One, 2009, 4(10), e7603.

[177] Zim MM, Sam IC, Omar SS, Chan YF, AbuBakar S, Kamarulzaman A, Chikungunya infection in Malaysia: Comparison with dengue infection in adults and predictors of persistent arthralgia, J Clin Virolo, 2013, 56(2), 141–45.

[178] Bates PA, Leishmania sand fly interaction: Progress and challenges, Curr Opin Microbiol, 2008, 11(4), 340–44.

[179] Naucke TJ, Schmitt C, Is leishmaniasis becoming endemic in Germany, Int J Med Microbiol, 2004, 293, 179–81.

[180] Perrotey S, Mahamdallie SS, Pesson B, Richardson KJ, Gállego Culleré M, Ready PD, Postglacial dispersal of Phlebotomus perniciosus into France, Parasite, 2005, 12, 283–91.

[181] Ready PD, Leishmaniasis emergence and climate change, Rev Sci Tech, 2008, 1, 27(2), 399–412.

[182] Nichol ST, Beaty BJ, Elliott RM, Goldbach R, Plyusnin A, 2005, Genus Phlebovirus, Virus Taxonomy: Eighth Report of the International Committee on Taxonomy of Viruses, Faguet CM, Mayo MA, Maniloff J, Desselberger U, Ball LA, eds., Elsevier Academic Press, United states, 709–11.

[183] Palacios G, Savji N, Da Rosa AT, Desai A, Sanchez-Seco MP, Guzman H, Lipkin WI, Tesh R, Characterization of the Salehabad virus species complex of the genus Phlebovirus (Bunyaviridae), J Gen Virol, 2013, 94(Pt 4), 837–42.

[184] Ergunay K, Sayiner AA, Litzba N, Lederer S, Charrel R, Kreher P, Us D, Niedrig M, Ozkul A, Hascelik G, Multicentre evaluation of central nervous system infections due to Flavi and Phleboviruses in Turkey, J Infect, 2012, 65(4), 343–49.

[185] Navarro-Marí JM, Gómez-Camarasa C, Pérez-Ruiz M, Sanbonmatsu-Gámez S, Pedrosa-Corral I, Jiménez-Valera M, Clinic–epidemiologic study of human infection by Granada virus, a new Phlebovirus within the sandfly fever Naples serocomplex, Am J Trop Med Hyg, 2013, 88(5), 1003–06.

[186] Kocak TZ, Weidmann M, Bulut C, Kinikli S, Hufert DG, Demiroz AP, Clinical and laboratory findings of a sandfly fever Turkey Virus outbreak in Ankara, J Infect, 2011, 63(5), 375–81.

[187] Coates BJ, 1963, Infectious diseases, Internal Medicine in World War II, Washington D.C., USA, Office of the Surgeon General Department of the Army, Washington DC, USA.

[188] Minodier P, Parola P, Cutaneous leishmaniasis treatment, Travel Med Infect Dis, 2007, 5(3), 150–58.

[189] Steere AC, Coburn J, Glickstein L, The emergence of Lyme disease, J Clin Invest, 2004, 113(8), 1093–101.

[190] Wormser GP, Dattwyler RJ, Shapiro ED, Halperin JJ, Steere AC, Klempner MS, Krause PJ, Bakken JS, Strle F, Stanek G, Bockenstedt L, The clinical assessment, treatment, and prevention of Lyme disease, human granulocytic anaplasmosis, and babesiosis: Clinical practice guidelines by the Infectious Diseases Society of America, Clin Infect Dis, 2006, 43(9), 1089–134.

[191] Blanton L, Keith B, Brzezinski W, Southern tick-associated rash illness: Erythema migrans is not always Lyme disease, South Med J, 2008, 101(7), 759–60.

[192] Institute of Medicine, US, 2011, Committee on Lyme Disease and Other Tick-Borne Disease: The State of the Science, Critical Needs and Gaps, Amelioration, and Resolution of Lyme and Other Tick-Borne Diseases: The Short-Term and Long-Term Outcomes, Washington (DC), National Academics Press (US).

Amit Kumar Singh

Chapter 5
Mycology

Abstract: Mycology is a very vast discipline of life science that deals with the study of fungi, their biochemical properties, environmental impact, pathogenesis and genetic analysis. Fungi are well known for their harmful as well as beneficial aspects in the area of microbiology. Fungi are responsible for a large number of diseases in humans and animals also. This chapter deals with the basics of mycology, classification of mycosis, diagnosis, microscopic examinations, advanced diagnosis and major fungi-mediated diseases.

Keywords: fungi, mycology, mycosis, pathogenesis, fungal detection, mycotoxins, mycotoxicoses

5.1 Introduction

Mycology is concerned with the discipline of life science that deals with the study of fungi, their biochemical properties, environmental impact, pathogenesis and genetic analysis. The term "mycology" is composed of two "Greek" words, where mukēs stands for "fungus" and logia means "study." These microorganisms are considered under a separate kingdom "Fungi," including filamentous fungi, molds, yeasts and mushrooms. Fungal strains are well known for their potential of producing commercial enzymes, food products, antibiotics, pigments and other valuable industrial products. Today, fungal strains are regularly used for various applications in different industries. Among the known fungal species, it is estimated that only 325 species are infectious to humans. The small subset of human pathogenic fungi is also plant pathogenic with the capability of cross-kingdom infection. These crossover pathogens include fungi, which belongs to phylum Ascomycota and subphylum Mucoromycotina. Most of ascomycetes crossover fungi are densely melanized, also called as dematiaceous fungi [1]. The fungal species with crossover pathogenesis are predominantly soil saprophytes with the abilities of causing infections in both plants and humans. Majority of fungi can establish nonfatal infection with limited but harmful effects on health, whereas they can cause serious and life-threatening infections in the immunocompromised individuals [2, 3]. This chapter covers the basics of fungal diseases, infections, fungal diagnosis, fungal toxins and antifungal agents.

Amit Kumar Singh, Immunology and Microbial Diseases Laboratory, Albany Medical Center, Albany, New York 12208, USA, Contact +15182588222, e-mail: singha5@amc.edu

https://doi.org/10.1515/9783110517736-005

5.2 Mycosis

The term mycosis is used for a disease caused by the fungal infection. Mycosis can be classified depending upon the type and degree of tissue involvement in the fungal infection. It can also be classified in the contest of host immune responses against fungal infection. In general, mycosis may be superficial, subcutaneous or systemic.

5.3 Classification of mycosis

Based on the site of infection, mycosis can be categorized into (1) superficial, (2) subcutaneous and (3) systemic.

5.3.1 Superficial

Superficial or cutaneous mycoses are fungal infections of dermatophytes on nail, hair shaft and skin (limited to the stratum corneum). The three most common types of superficial mycosis are infections of dermatophytes such as *Trichosporon beigelii* (white piedra) and *Piedraia hortae* (black piedra), *Malassezia furfur* (pityriasis versicolor) and *Phaeoannellomyces werneckii* (*Tinea nigra*) and non-dermatophytic molds [4, 5]. Pityriasis versicolor is a superficial mycosis with hypo- or hyper-pigmentation of the skin. The superficial infection of *Malassezia furfur* at keratin layer of skin causes pityriasis versicolor, dandruff/seborrheic dermatitis and folliculitis [6]. Black piedra is a mycosis caused by the infection of *Piedraia hortae* on the hair shaft. In contrast, white piedra is the infection of *Trichosporon beigelii* to the hair shaft characterized by numerous, soft, discrete, concretions or nodules. White piedra infection acquired by an immunocompromised individual can turn out into clinical conditions. *T. nigra* is a superficial mycosis, which typically presents as a brown to black silver nitrate-like patches on the palm of hands or sole of feet. It is caused by the parasitic infection of pigment yeast, *Phaeoannello myceswerneckii*. *Candida albicans* is the causative agent of candidiasis or "thrush" in humans [7]. It retains commensal relation and inhabits as normal flora of the gastrointestinal tract and vagina. However, during impaired immunity or collapsed health condition, the organism aggressively multiplies and causes disease [8, 9]. Superficial mycoses may also be classified as dermatophytoses or dermatomycoses caused by agents of the genera *Epidermophyton*, *Microsporum* and *Trichophyton* [9, 10]. The dermatophytoses are characterized by rashes, scaling and itching at the anatomic site. For example, *Microsporum* is a dermatophyte fungus of cats and dog, which can be transmitted to human and can cause tinea capitis (lesions on head) or tinea corporis (glabrous skin lesion). Other dermatophytoses species usually infect on the hair and skin but do not involve nail infections, i.e., *Trichophyton* species. Whereas *Epidermophyton floccosum* infection usually occurs only at the skin and nails but do not infect hair shafts and follicles [9, 10].

5.3.2 Subcutaneous mycoses

Subcutaneous mycoses can be fostered by the heterogeneous group of fungi that develop infection into the skin, subcutaneous tissues and organs of host. Majority of subcutaneous mycosis caused by the infections of fungi is responsible for chromoblastomycosis, lobomycosis, mycetoma, rhinosporidiosis, sporotrichosis, zygomycosis and phaeohyphomycosis [11–13]. Chromoblastomycosis is the infection of dematiaceous fungi, characterized by verrucoid lesions of the skin. Infection is usually limited to the subcutaneous tissues without affecting bones, tendons or muscles. *Fonsecaea pedrosoi*, *F. compacta*, *Cladosporium carrionii* and *Phialophora verrucosa* are commonly known as causative agents of chromoblastomycosis [12]. In contrast, mycetoma is a subcutaneous mycosis that damages contiguous bones, tendons and skeletal muscles. Mycetoma infections are caused as greater diverse fungal infections and can be classified as eumycotic and actinomycotic mycetoma [13]. Several mycetoma causing fungal species are densely pigmented and leads brown to black color appearances. Such fungi are considered as dematiaceous (melanized) fungi, causing a wide range of infections from superficial to deep tissue infection. Sporotrichosis is the third general class of subcutaneous mycoses involved in the infection at the traumatic inoculation points [11, 14].

5.3.3 Systemic or deep tissue mycoses

Systemic or deep tissue mycoses are the chronic granulomatous disease caused by primary and/or opportunistic fungal pathogens. *Histoplasma capsulatum*, *Coccidioides immitis*, *Blastomyces dermatitidis* and *Paracoccidioides brasiliensis* are primary pathogenic fungi, which can cause mild or clinical infection in normal patients [15, 16]. Whereas opportunistic pathogens include *Candida* species, *Aspergillus* species, *Fusarium* species, *Penicillium marneffei*, *Cryptococcus neoformans* and *Trichosporon beigelii*, require an immunocompromised host to establish infection [17, 18]. Patients with chronic diseases/infections (i.e., cancer, organ transplantation, surgery and AIDS) are at higher risk. The life-threatening primary deep tissue mycosis is established with the inhalation of high fungal inoculum while living in the endemic region or traveling through. Whereas infection by opportunistic fungi occurs mostly in immunocompromised patients via the respiratory tract, alimentary tract or intravascular devices.

5.3.3.1 Primary mycosis

Primary mycosis in deep tissues is mostly asymptomatic. Mild clinical or subclinical infections can establish spherules in lungs with the inhalation of arthroconidia from *Coccidioides immitis* called as coccidioidomycosis. The disease manifests to other vital organs, including brain, liver and bones. The infection may be life-threatening and require prolong treatment with aggressive antibiotic doses. Similarly, histoplasmosis infection caused by inhalation of *H. capsulatum* conidia that convert *vivo* into the blastoconidia and disseminate to the mediastinal lymph nodes, liver, spleen, bone marrow and other vital organs may result in a life-threatening infection in the immunocompromised person [16]. Blastomycosis is also a primary pulmonary infection caused due to inhalation of conidia from *B. dermatitidis* [19, 20]. The clinical pattern of blastomycosis is marked with chronic pulmonary infection, fibrotic reactions and dissemination to other vital organs, including bone marrow and male genital organs.

5.3.3.2 Opportunistic mycosis

Opportunistic mycoses are infections of inherently lesser virulent fungal species, mostly noninfectious or latent asymptomatic to the healthy host. Under the impact of immunosuppressive agents, immunomodulator or chronic infection resulted in immunocompromised conditions that led to the invasion of common fungal pathogens such as *Aspergillus* species, *Candida* species, *Cryptococcus* species, *Fusarium* species and *Mucor* species at superficial or deep tissues in the host. Superficial candidiasis may occasionally involve with the infection at mucosal and epidermal surfaces of lung, intestines, urinary bladder and vagina. Deep or visceral candidiasis is caused by infection in the brain, eyes, kidney, spleen and livers [7, 21]. The portal of entry for aspergillosis is the respiratory tract; however, it may also enter the host via compromised skin epithelium. Aspergillosis infection established in the lungs and paranasal sinuses usually disseminate to the brain, kidneys, liver, heart and bones. Members of class Zygomycetes (*Rhizopus*, *Rhizomucor* and *Mucor* species) may cause invasive sinopulmonary infections called zygomycosis. Mucormycosis or rhinocerebral syndrome is a life-threatening zygomycosis prevalent in diabetic patients [22, 23]. Cryptococcosis is a pulmonary infection caused due to the infestation of *Cryptococcus* species. This is typically an opportunistic fungal pathogen, which develops pneumonia and/or meningitis in patients with defective immunity [21, 24]. Phaeohyphomycosis is cutaneous, superficial and deep tissue infection of pigmented fungi, especially in the brain. Incidence of subcutaneous or deep tissue infection by saprophytic fungi with hyaline hyphal elements (*Fusarium* species) may develop hyalo hyphomycosis [18]. In patients with depleted neutrophil or defective neutrophil function, subcutaneous infection by saprophytic fungi may cause pneumonia, fungemia and disseminated infection with cutaneous lesions.

5.4 Laboratory diagnosis of fungal diseases

The progressive increase in the frequency of invasive fungal infection cases notably emphasizes for the development of quick and reliable diagnosis tools. Conventional culture methods are considered gold standard for the determination of exact etiology of fungal infection; however, it usually takes 2–3 weeks of time and may fail against noncultural fungal strains. The limitation can be overcome with the application of nonculture-based diagnosis methods, including microscopic examination, antigen–antibody interaction, mass spectroscopy, nucleic acid-based detection methods and detection of fungal metabolites. Early and precise identification of causative agent is one of the important challenges of therapeutic interventions against fungal pathogens. Microscopic examination is the basic but important part to examine fungal mycelium. Microscopic visualization of fungus can be improved by the use of various available straining methods [25, 26].

5.4.1 *In vitro* culture test

In vitro culture persists as a crucial method for the purpose of isolation and diagnosis of fungi from environmental, clinical samples and food samples using a battery of media. The culture media can be categorized into primary recovery nonselective media like "brain heart infusion agar," "potato dextrose agar" and "Sabouraud's heart infusion" media followed by selective media that are specifically tailored to isolate specific fungi of interest. Currently, there are no single (specific) medium or a combination/mixture of media that is suitable for all samples or specimens. Media selection should be conceded precisely, depending upon the type of specimen and suspected fungal agents (Table 5.1). In case of nonsterile environmental, food, water and clinical samples, it is vital to supplement primary nonspecific media with broad-spectrum antibacterial agents such as chloramphenicol and gentamicin to prevent unwarranted bacterial contaminations and cycloheximide to inhibit most saprobic molds, whereas samples obtained from a sterile site can be processed without any antibacterial agents. Some of the recent findings have identified the inhibitory effects of antibiotics on various fungal species such as *Aspergillus* species, *Scopulariopsis*, *Pseudallescheria*, *Fusarium*, *Trichosporon*, some *Talaromyces* (*Penicillium*) *marneffei*, mucoraceous fungi, few dematiaceous (fungi) and yeasts such as *Candida* sp. and *Cryptococcus* sp. are found with suboptimal/inhibition of growth in the presence of cycloheximide in media [26]. Hence, it is imperative to keep a set of primary inoculation of specimen with and without inhibitory agents (Table 5.1).

Table 5.1: Culture methods and media for selective identification of fungi from clinical and environmental samples.

Media	Indicator of use
Brain heart infusion agar with antibiotics	Primary recovery of pathogenic fungi is exclusive for dermatophytes. Antibiotics act as a selective agent for growing microorganism.
Caffeic acid agar	*Cryptococcus neoformans* produces melanin, resulting in black colonies.
Chromogenic agar	Isolation and differentiation of urinary tract pathogens. Easy to differentiate mixed flora on CHROMagar™ Orientation.
Christensen's urea agar	Also known as urease agar medium require the identification of urease-producing members of genus *Proteus*.
Czapek's agar	Growth medium for fungi, useful for the differential identification of *Aspergillus* species.
Dermatophyte test medium	Primary recovery of dermatophytes; recommended as a screening medium only
Inhibitory mold agar	A selective medium used for the isolation and cultivation of *Histoplasma capsulatum*, dermatophytes and other pathogenic fungi from clinical and nonclinical specimens.
Malt extract agar	Malt extract agar is the formulation recommended by Thom and Church for the growth and isolation of mold and yeasts.
Mycosel	Mycosel agar is a highly selective medium for the isolation of pathogenic fungi (dermatophytes). Good for cultivation of mushrooms.
Niger seed agar	Niger seed agar (Staib agar) is a selective agar for isolation and identification of *Candida dubliniensis* from clinical samples. Useful in *Cryptococcus neoformans* complex identification.
Nitrate reduction medium	Detection of *Cryptococcus* species.
Sabouraud's dextrose agar	Primarily used for the isolation of dermatophytes, other fungi and yeasts
Yeast extract phosphate agar	Used for the isolation of dimorphic fungi, *Blastomyces dermatitidis* and *Histoplasma capsulatum*.

5.4.2 Microscopic examination

Microscopic examination of the clinical, environmental and food samples for the identification of fungal specimen is one of the rapid and most accurate modes of diagnosis. In general, samples with adequate quantity are directly mounted on the microscopic slides and assess the colorless, whitish or pale yellow-colored vegetative structures, critical for correct identification of causal fungal agents, whereas a little amount of specimen inoculated in appropriate media enhances recovery and identification of fungus. The examination of fungal specimens can be further improved by numerous staining methods (Table 5.2), employed to improve visualization of hyphae, conidia and other fungal structures under light microscope [25, 26]. Besides standard light or phase-contrast microscopy, transmission electronic microscopy enables observation and analysis of fine structural details of fungi. The morphological characteristics of fungi in terms of conidia type, shape, size, color of fungal hyphae, texture and arrangement details can help in the instant identification of fungal contaminants, however, requires good knowledge and experience in the subject [26].

Table 5.2: Staining techniques for the identification of fungi.

Staining methods	Applications and limitations
Acid-fast staining	Ziehl–Neelsen stain or acid-fast stain for *Nocardia* species and other aerobic *Actinomyces*. Strong background in staining of clinical samples makes difficult to observe.
Auramine rhodamine stain	Also called as Truant auramine–rhodamine stain. Similar to Zeil–Neelsen stain, it stains *Nocardia* species, however, observed nonspecific reactivities.
Calcofluor whitestain	Binds to the chitin in the cell walls of the budding yeast or fungi. Requires expensive fluorescent microscope, and in the presence of various biological fluids and component, it emits strong autofluorescence background.
India ink stain	Useful for the *Cryptococcus* species containing large polysaccharide against India ink black background. The absence of discernible capsule in some strains of *C. neoformans* make it fewer than 50% positive.
Lactophenol cotton blue	The most popular for mounting and staining yeast and molds. Cotton blue is an acid dye that stains the chitin present in the cell walls of fungi. Require mechanical method to dislodge fungal components.
Masson-Fontana stain	The Fontana-Masson stain is intended for use in histopathological visualization of melanin pigments present in cell wall of dematiaceous fungi. It is also evident that the nonspecific reactivity with some non-dematiaceous fungi including Zygomycetes, *Aspergillus* and *Fusarium* spp.

Table 5.2 (continued)

Staining methods	Applications and limitations
Methenamine silver stain	Grocott-Gomori's stain is a popular histological staining of fungi. Also shows reactivity with yeast-like fungus which causes a form of pneumonia called *Pneumocystis pneumonia*.
Papanicolaou stain	Papanicolaou or Pas is a cytological stain for various bodily secretions that stains fungal elements from pink to blue.
Periodic acid-Schiff (PAS) stain	The PAS procedure is mostly opted by histologic laboratories to study mycotic infections. *Coccidioides immitis*, *Blastomyces* species and other fungi with prominent carbohydrates shows strong reactivity. Nocardia species do not stain well. Stain shows non-specific reactivity with several gram-negative bacteria and *Mycobacterium* species.
Potassium hydroxide	Quick, inexpensive fungal test for dermatophytes. Clearing of specimen to make fungi more readily visible. Need optimization of experiment with different fungal species and may develop background artifacts.
Wright's stain	Most often used to detect *H. capsulatum* and *C. neoformans* complex in disseminated disease.

5.4.3 Serodiagnosis

The immunologic approach to diagnose invasive fungal infection is carried out by detecting genus-, species- or strain-specific antigens in host body fluids or other environmental samples. Target antigen should be conserved and present sufficiently within the fungal species of interest. A large number of studies have focused on using fungal cell wall components for the development of specific antibodies as an antigenic marker [27–29]. Whereas, a 47-kDa and a 48-kDa cytoplasmic proteins were selected as target antigens from Candida species [30]. These novel cytoplasmic markers are utilized for the identification of systemic or deep tissue candidiasis [31, 32]. Polysera generated against fungal surface or cytoplasmic antigens were usually found highly cross-reactive with various host antigens or other closely related pathogenic fungi [29, 31]. Along with, the application of polysera for serodiagnosis are facing setbacks like batch-to-batch variations, cross reactivities and chances of zoonotic transmission of disease [33]. The application of monoclonal antibodies for the immunodiagnostic assays retain several advantages over polysera such as consistency in batch productions, reduced cross-reactivity and chances of cross-infection.

5.4.4 PCR detection method

Polymerase chain reaction (PCR) detection method offers several attributes to over-come the shortcoming of conventional techniques of fungal diagnosis. A combination of specific primers and probes designed for the target gene can further enhance a specific diagnosis of genus or species of fungi using a fragment sequencing. Along with, the recent advancement in the real-time PCR platforms enable quantification of fungi in the food, environmental and clinical specimen. The quantification of fungal load in clinical samples may provide information about burden or progression of disease in patients. Multiplexed PCR or real-time PCR assays involved in a combination of different sets of compatible primer probes enable the simultaneous detection of more than one pathogenic fungus. Procedural, exogenous or cross-fungal contaminations are critical while applying highly sensitive multiplexed PCR reactions that further compounded with the cross-reactivity of primers or probes [34, 35]. Restriction fragment length polymorphism and random amplified polymorphic DNA assessment of the fungal genome show promising results with gene locus specific to subtyping in PCR and sequence comparisons of targets [38]. The analyses of specific gene target and sequence variations are being utilized in detection and genotyping in a single PCR and subsequently performed sequencing [29].

5.4.5 Sequencing-based detection

Determination of DNA and RNA sequence is the most revolutionized areas of bio-medical research, utilized in the identification of various fungal species [36]. Recent advent in the sequencing of DNA fragment and high-throughput DNA sequencing enables identification of fungi in food, environmental and clinical samples. The variable DNA sequence and fragment length in the areas of intervening internal transcribed spacer (ITS) regions of the nuclear ribosomal repeat unit called ITS1 and ITS2 regions have a critical role in the development of functional rRNA. Sanger sequencing of ITS1 and ITS2 regions among species show variation in nucleotide stretch and act as a signature region for molecular detection of fungi. Along with the sequence homology within the ribosomal rDNA genes of fungi (18S, 5.8S and 28S genes) and differences within the spacer regions (ITS1 and ITS2) are the genetic basis for the organization of the fungi into taxonomic groups [36, 37]. In the advent of second- and third-generation high-throughput sequencing technologies, hundreds of thousands of limited-length sequences are generated with adequate information for differential identification of fungi at species level [36, 38].

5.4.6 MALDI analysis

The presence of inherent biological complexity and different phenotypes during various stages of life/growth makes the fungal identification a challenging task [39]. In addition, recent identification of several nonculturable fungi further deepens this problem and emphasizes on thrust for better reliable molecular techniques. Matrix-assisted laser desorption/ionization time-of-flight mass spectrometry (MALDI-TOF MS) has been identified as one of the most reliable means of identification of various microorganisms. Large number of fungal species and genera, i.e., *Aspergillus*, *Fusarium*, *Penicillium* or *Trichoderma* and yeasts obtained from clinical isolates (e.g., *Candida albicans*), were characterized and classified using MALDI-TOF MS. Intact cell/spore or extracted surface proteins can be selected to perform MALDI-TOF MS, where it can be identified via direct acquisition of mass spectra. The application of MALDI-TOF enables evaluation and identification of specific markers/ions, which can be utilized in differential identification of microorganism at genus, species or strain level by using software and spectral database of reference strains [40]. Till now, there are no versatile methods/protocols for MALDI-TOF MS of fungal microorganisms that have been established. MS may fail to identify rare species in the absence of reference spectrum in database.

5.5 Dimorphism in the pathogenic fungi

Dimorphism of fungi is the identification of physiological and morphological changes that occur in a fungal strain, which led to conversion of fungal hyphal form to a yeast-like form under the impact of environmental conditions. The dimorphic fungi capable of causing systemic disease belong to the order Onygenales, including genera *Blastomyces*, *Histoplasma* and *Emmonsia* under Ajellomycetaceae and Onygenaceae families [41]. The Onygenales are known to share various characteristics (general): their sexual stages, i.e., teleomorphs, form rudimentary asci that are surrounded by a hyphae network. Also, their asexual (anamorph) species generally possess one of the two forms; either aleurioconidia or arthroconidia in chain form having alternate viable and nonviable cells [41, 42]. Molecular studies have identified genus *Paracoccidioides* with *P. brasiliensis* and *P. lutzii* under human fungal pathogen clade belonging to family Ajellomycetaceae (Ascomycetes); however, teleomorph of the *P. brasiliensis* has yet to be discovered [43].

5.6 Histoplasmosis

Histoplasmosis is caused by the intranasal inoculation of pathogenic fungi called *Histoplasma*. It is commonly transmitted by birds and bats by their dropping with fungal spores that often become airborne during cleanup projects and other anthropogenic activities. After 1–3 weeks of incubation period, the host develops lung disease similar to pneumonia. Upon re-exposure, this incubation period turns out to be shorter for 4–7 days. Histoplasmosis is very rarely contagious and has been reported only during organ transplantation. Dissemination most commonly occurs among the individuals with immunocompromised conditions, including HIV-positive patients and transplant recipients [44, 45]. The thermally dimorphic status of *H. capsulatum* is demonstrated by a filamentous mold condition in environmental condition; and at temperature lower than 35 °C and yeast phase in tissue; and at more than 35 °C temperatures. This dimorphism of fungi offers best growth in nitrogen-enriched soils, particularly those areas having high numbers of birds or bat guano. Birds that are dropping *H. capsulatum* are not colonized or infected (because of higher body temperatures), and the droppings of birds are primarily a source of nutrient for the fungi. These soil contaminants can persist for up to 10 years from the date of rooster clearance [44]. North America has been listed as one of the most histoplasmosis-endemic regions, which is now spreading to central and southern part of the country [46]. The endemicity of disease also includes most of Australia, Africa and also eastern Asia, particularly Malaysia and India.

5.7 Blastomycosis

Blastomycosis is a systemic fungal disease of human and other animal cause by the infection of thermally dimorphic fungi *Blastomyces*. The pathogen naturally inhabits in the soil, particularly moist-acidic soils near waterways that have high level of nitrogen and also organic content. It converts in the yeast phase from the mold phase under specific appropriate condition, and the airways inhalation of spores is causing chest pain, difficulty in breathing, fever and chills. Skin infection begins as very small bumps and may be established into warty patches surrounded by tiny, painless collections of pus [47]. Patients with severe blastomycosis are treated with intravenous administration of amphotericin B, and must be continued for 6–12 months. Without treatment, blastomycosis slowly worsens but rarely leads to death [48].

5.8 Coccidioidomycosis

Coccidioidomycosis or valley fever occurs due to inhalation of the fungal spores of *Coccidioides* species. This infection is predominant in southern Arizona, California, southern New Mexico, west of Texas and Brazil. Visitors or travelers entering in these endemic regions are prevalent of developing infection, especially in population retaining immunocompromised health conditions [49]. The pathogen contrives dimorphic condition by sustaining mold-producing septate hyphae and arthroconidia in the environmental condition and in culture at room temperature. Inside the host and *in vitro* growth on specific media allow transformation of arthroconidia into a spherule contriving hundreds to thousands of endospores. The virulent spores usually aerosolized by anthropogenic activities and infection may develop into flu-like symptoms with fever, shortness of breath and headache within 1–2 weeks post-exposure. These symptoms usually go away on their own after weeks to months; however, 3–6 months of treatment with fluconazole or another type of antifungal medication has been recommended. In case of severe lung infections or infections that have manifested to other vital organs, aggressive antifungal treatment is needed for longer than 6 months [49, 50].

5.9 *Emmonsia* disease

Emmonsia disease is a rare systemic mycosis in humans. The pulmonary infection of *Emmonsia* is established with the inhalation of aerosolized conidia, released from mycelia present in the soil. Inside the host's lung, they elicit granulomatous reactions with wide range of diseases from subclinical pneumonia to respiratory failure and, occasionally, death. The severity of diseases largely depends on the size of inoculum [51].

Among the three known genus, *Emmonsia* includes three species, i.e., *E. crescens*, *E. parva* and *E. pasteuriana*. *E. pasteuriana* is relatively more prevalent in South African region. *E. crescens* is known to cause adiaspiromycosis in more than 118 mammalian species, globally. The human case of adiaspiromycosis is very rare since the first human case was reported in 1964. In the environmental condition and in vitro culture at room temperature, *Emmonsia* species can be characterized as a mold capable of producing single-celled conidia of size ~4 μm. The conidia of both *E. parva* and *E. crescens* transform into adiaspores inside the host similar to the spherules of the *Coccidioides* species [51, 52].

5.10 Fungal and fungal-like infections of unusual or uncertain etiology

Various clinical and histopathological studies are describing few exotic and unusual infections in human population with characteristics of fungal or pseudofungal infections. These include nonreplicative pulmonary infection (adiaspiromycosis), two subcutaneous mycoses (entomophthoromycosis and lobomycosis), two algal infections (chlorellosis and prototothecosis) and an infection due to the oomycete *Pythium insidiosum*. The identification of these rare infections is based on detection of characteristic symptoms, histopathology and structures.

5.10.1 Adiaspiromycosis

Adiaspiromycosis (also known as adiaspirosis or haplomycosis) is a fungal infection which affects lower order mammals including rodents and are rarely infective in human population [53]. Infection usually develops due to inhalation of spores from fungi *Emmonsia* species. Rodents have been considered as a natural reservoir for this zoonotic disease, although the zoonotic transmission of a disease from rodents to humans is very rare [54, 55]. The pulmonary infections of *E. crescens* are localized and asymptomatic with lesions produced by the enlarging conidia with discrete or confluent fibrotic granulomas, each containing one or more spherical, thick-walled adiaconidia, significantly large that can be seen with the naked eye [51].

5.10.2 Coccidioidomycosis

Coccidioidomycosis or Valley fever is a systemic mycosis caused by inhalation of aerosolized spores of fungus *Coccidioides* [56–59]. This saprophytic fungal infection is mostly on their own within weeks to months, without any medical treatment. However, patients cured with the immunocompromised health conditions develop a wide range of clinical disease from a mild pulmonary pneumonia up to respiratory failure [58–60]. In recent years, the increasing population in southern Arizona and central California causes soaring hike in the numbers of infections per year up to ~150,000 [57–59].

5.10.3 Entomophthoromycosis

Entomophthoromycosis (or entomophthoramycosis) is caused by fungi belonging to the Entomophthorales, including basidiobolomycosis and conidiobolomycosis [61, 62]. Entomophothoromycosis is a rare fungal disease in humans or other animals caused by the fungi of order Entomophthorales, including two genera *Basidiobolus* and *Conidiobolus* that are infectious to humans. The entomophthoromycosis has predominant infection in the region with higher percentage of adipose tissues since organisms thrive largely on fatty substances. The chronic subcutaneous infection (basidiobolomycosis) caused by *B. ranarum* has been reported in different parts of Africa, India, Indonesia and Southeast Asia, where fungal implantation has been identified into subcutaneous tissues of thighs, buttocks or trunk [22, 62, 63]. Painless mobile disk-shaped, rubbery masses with inflammatory cellular materials usually develop into seldom life-threatening erythematous plaques. Additionally, skin ulcer and lymph node enlargement may be observed. In contrast, the human pathogenic *Conidiobolus* are aerosol transmitted and infect nasal sinuses and soft facial tissues. The conidiobolomycosis establishes a slow extending painless, firm swelling into the nasal bridge, upper and lower face including orbit [22, 62]. The differential diagnosis of entomophthoromycosis has been proffered on biopsy in the tissue section and histopathological analysis. Systemic prolonged administration of broad-spectrum antifungal agents (i.e., amphotericin B, imidazole and potassium iodide) individually or cocktail mixture in combination with surgical debridement is recommended for the successful treatment.

5.10.4 Lacaziosis

Lacaziosis, also known as lobomycosis, is a self-limited chronic fungal infection of the skin. The disease is endemic in southern and central parts of America. Natives of the Brazilian rain forest call this disease "miraip" or "piraip," meaning "that burns" [64]. The disease is caused by a yeast-like fungus when grown in aquatic environment. Group of people received traumatic implantation, arthropod sting, snakebite, stingray sting and wound acquired while cutting vegetation usually get this fungal infection [65]. Biopsy and histopathology demonstrated characteristic yeast cells in lesion exudates or tissue sections of patients with *L. loboi* infection. The natural reservoir of *Lacazia loboi* is not completely elucidated while soil and vegetation seem to be likely sources of infection. *Lacazia loboi* has also been recovered from lobomycotic lesions of the Guiana dolphin *Sotalia guianensis* from the Surinam River estuary and *Tursiops truncatus* dolphins [65–67]. It often develops at sites of minor trauma, but sometimes, no history of trauma can be recalled. The lateral transmission of disease from human to human has never been reported, although occupational exposure of a person to the infected dolphin may cause infection. Surgical excision has been considered as the best treatment; however, long-term application of clofazimine is recommended for the postsurgical application to prevent relapse of disease [68, 69].

5.10.5 Rhinosporidiosis

Rhinosporidium seeberi is known as the causative agent of an enigmatic disease called rhinosporidiosis. It is a noncontagious, chronic granulomatous infection in humans and large domestic animals, characterized by the production of large polyps with high recurrence rate of infection. The fungal infestation primarily affects the nasal cavity, conjunctiva and urethral mucosa linings where they grow into a polyploid mass that affects mucosa and internal cavity by causing congestion and inflammatory responses [70]. It is considered in the class Mesomycetozoa family Rhinosporideacae as each circular cell is converted into a single sporangium. The disease spreads globally with prevalence in more than 70 nations at a diverse geographical distribution and clinical features. Although the disease is evident sporadically in parts of Europe, Africa, southern United States, South Africa and South American temperate regions, and western and middle eastern countries, however, it has endemic status in many Asian countries, especially tropical regions, India, Sri Lanka, Bangladesh, Pakistan, Brazil and Uganda [71, 72]. *Rhinosporidium seebri* spreads through trauma that occurs during accident or surgery and causes spread of organism to various regions of the body. The infection is mainly seen in the respiratory tract infection. Reports of *R. seeberi* dissemination to the distant organs have been recorded but the exact immune mechanism in humans is still unclear. A number of groups suggested that the thick outer wall contriving antigenic structures inside the cytoplasmic space bring lesser chances of reactivity with specific antibodies inside the host [72, 73].

Despite the first identification of rhinosporidiosis was back in 1892, large enigma persists around this microorganism due to the absence of a proper *in vitro* culture media and an established animal model. The crisis continues with the inability to identify its natural reservoir. Water was claimed as natural reservoir, since the presence of spherical bodies that are compatible with endospores or juvenile sporangia of *R. seeberi* has been reported by earlier several groups, but a recent finding based on PCR techniques was failed to confirm it. Surgical removal or nasopharyngeal polyps by cold and host snare is in practice to treat rhinosporidiosis. Cryosurgical excision of lesion from the nasopharyngeal cavity checks the recurrence and further spread of infection. Morbidity is low and generally due to secondary bacterial infection of the lesion [62, 74].

5.11 Mycotoxins and mycotoxicoses

Mycotoxins are toxic secondary metabolites produced by various fungal species under favorable conditions. Dietary, respiratory and epidermal exposure of these fungal toxic metabolites can cause disease condition to the host, collectively called as mycotoxicosis. Mycotoxicosis can create a range of diseases from mild uneasiness or annoying to life-threatening poisoning in the host. In nature, these chemical components can act as a protective agent for producing fungi against other microorganisms, host and predators. Polyketides (like fatty acids) are produced using acyl coenzyme A (acyl-CoA), alkaloids from the prenylated aromatic amino acids, nonribosomal peptides (from amino acids) and also terpenes from isoprene. Till now, more than 200 types of mycotoxins have been studied; however, the field of mycotoxins is considered in a primitive stage of study and experts believe that majority of mycotoxins are yet to be discovered. Aflatoxin clusters were the first mycotoxin gene clusters to be characterized in both *Aspergillus flavus* and *A. parasiticus*.

5.11.1 Aflatoxins

Aflatoxins are the first mycotoxins discovered and studied; mycotoxins are produced by toxic molds such as *Aspergillus* species, *Penicillium*, *Rhizopus*, *Mucor* and *Streptomyces* [75]. Chemically, aflatoxins are difuranocoumarin derivatives that generate many strains of *A. flavus* and *A. parasiticus*. Among the four known major aflatoxins are B1, B2, G1 and M1 [76]. Aflatoxin B1 is the most abundant, most studied and potent natural carcinogen. Contamination of food with aflatoxins may cause various economical and health-related issues and substantially degrade the export value of food grains [77]. The most remarkable outbreak of aflatoxicosis (in humans) was found in Kenya (2004), followed by outbreaks in 2005 and 2006 at the same region [78].

5.11.2 Citrinin

Citrinin is a low-molecular-weight mycotoxin originally isolated from *Penicillium citrinum*. It is an acidic lemon-yellow crystal, capable of forming chelate complexes. Citrinin is most commonly produced by the different strains of *Aspergillus* species, *Penicillium* species and *Monascus* species. The production of citrinin is influenced by various parameters, including nutritional factors such as oxygen, carbon and nitrogen sources, fatty acids and environmental factors, i.e., humidity and temperature. It possesses antibiotic, antifungal and antiprotozoal properties. Renal macromolecular production and mitochondrial function are largely affected and lead to cell death. The occurrence of this highly thermostable toxin in cereals, fruits, meat and their products develop threat of food poisoning. Till now, no human citrinin poisoning case has been reported, and its effect on human health is still unknown.

5.11.3 Cyclopiazonic acid

Cyclopiazonic acid is an indole-tetramic acid produced as a secondary metabolite by different species of *Aspergillus* and *Penicillium* [79]. The toxic effects of this mycotoxin have been analyzed by using *in vitro* animal cell monolayers and *in vivo* experimental animal models, including mice, rats [80], chickens [81], dogs [82] and pigs [83]. Intraperitoneal administration of cyclopiazoic acid sublethal dose in mice and rabbit produces hypothermia, hypokinesia, catalepsy, ptosis, sedation without loss of righting reflex, tremor, gait disturbance, dyspnea, opisthotonus, atypical convulsion and prolonged barbiturate-induced sleep. With tonic extension of toxin to the hind legs, respiratory arrest, cardiac fibrillation and epileptiform, ultimately achieve death of animal, though the contamination of cyclopiazoic acid has been traced in different food products, including grains, cheese, and meats. But no human outbreak has been recorded so far [84].

5.11.4 Ergot alkaloids

The fungal toxin ergot alkaloid is synthesized and secreted by almost all members of *Claviceps* genus. *Clavicep spurpurea* in sclerotia (ergots) growing on rye has been recognized as the agent responsible for causing ergotism or St. Anthony's fire. Other sources of ergot alkaloids are grasses infected with different endophytes and *Claviceps*. Different species of genus *Penicillium* and *Aspergillus* fungi are also known to produce these alkaloids. Ergotism in humans is usually established as gangrenous or convulsive subsequently ingestion of contaminated food. The infections are most prevalent in rye [85] and triticale that have open florets; however, small grains including wheat are considered as potential hosts. Ergotism has been characterized by the gangrene development, convulsions and also by abortion in sheep, cattle, and pigs, whereas symptoms in chickens resemble humans. In farm animals, ergotism has been reported for animals that consume grasses and feeds already contaminated with fungi (*Claviceps*, *Penicillium* and *Aspergillus*) and endophytes (*Acremonium coenophialum*). Ergot alkaloid is largely classified into two different indole alkaloid groups (clavines and lysergic acid) derived from tetracyclic ergoline. The clavinet alkaloid is predominantly produced by *Aspergillus*, *Balansia*, *Claviceps*, *Epichloë* and *Neotyphodium* spp. [86], while lysergic acid is the precursor for a wide range of ergoline alkaloids that are produced by the ergot fungus. Patients exposed to lysergic acid diethylamide produce hallucination and distort perception of reality. In modern days, ergotism is rare due to removal of these alkaloids during normal cleaning, baking and cooking of grains and vegetables.

5.11.5 Fumonisins

Fumonisins are among the recently discovered mycotoxins, first isolated in 1988 from *F. verticillioides* in Tygerberg, South Africa. They are shown to cause Equine leukoencephalomalacia (ELEM) in horses [87], porcine pulmonary edema syndrome in pigs [88] and liver cancer in rats [89]. The toxin producing *F. verticilliodes* fungi was also prevalently isolated from people consumed the home-grown corn in area with high incidents of esophageal cancer. Fumonisin toxins are also known to be produced by different *Fusarium* species and *Alternaria alternate*. They are considerably heat stable, need to remove from food chain by several processing, including wet washing of grains. Once enter inside the body, fumonisin blocks the synthesis of sphingomyelin, glycolipids and gangliosides which constitute important components of cell membrane and membrane-bound receptor for signaling between cells. Hence, the presence of fumonisin in food is associated with a wide range of disorder in the common aspect of membrane integrity and cell signaling. The daily tolerance intake of fumonisin in a person can be maximum 2 µg/kg body weight. However, majority of national permissive index for fumonisin in food grains is lower except southern Africa.

5.11.6 Ochratoxins

Ochratoxins have been described as one of the first groups of fungal metabolite that is toxic to animals. They are produced by *Aspergillus* species and few *Penicillium* species including *P. carbonarius* and *P. verrucosum*. Ochratoxin A is the most toxic among the group with observed carcinogenic, immunosuppressive, nephrotoxic and neurotoxic properties. Insertion of this toxin in the food chain by contaminating barley, oats, rye, wheat and coffee beans creates heavy socioeconomical impact by affecting livestock, dairy, meat and other industries [90]. A large concern of global toxin research has been pulled toward ochratoxin A as it can be passed in the food chain through milk and the pork. Hence, products derived from these industries are the major contributors to the human ochratoxin A exposure. The ethyl ester for of ochratoxin also known as ochratoxin C, chlorine atom devoid on C5 of the dihydromethylisocoumarin ring system, called ochratoxin B and other derivatives are comparatively lesser toxic with inferior economic values [91, 92].

5.11.7 Patulin

Patulin is a water-soluble lactone, first isolated as antibiotic component from *Penicillium griseofulvum* and *P. expansum* [93]. The simultaneous discovery of these compounds by various research groups historically labeled it as clavacin by Anslow and coworkers (1943), expansine by Van Luijk (1938), claviformin by Chain and coworkers (1942), clavatin by Bergel and coworkers (1943), gigantic acid by Philpot (1943) and myocin C by DeRosnay and coworkers (1952) [94, 95]. The initial application of patulin as a broad-spectrum antifungal and later expanded as an efficient antibacterial compound effective against more than 75 different Gram-positive and Gram-negative bacterial species. It was also suggested to be utilized as an effective drug against nasal congestion and common cold in humans. However, progressive clinical studies have suggested that patulin not only inhibits fungal and bacterial growth but also develops toxic effects to the animals and higher plants including cucumber, wheat, peas, corn and flax. Thus, in the late 1960s, patulin was considered as a mycotoxin, commonly present in juices made from *P. expansum*-contaminated fruits [96]. Although patulin is toxic at high concentrations, no natural patulin poisoning case has been reported in humans [95, 96].

5.11.8 Trichothecenes

Trichothecene are one of the major classes of mycotoxins synthesized as secondary metabolites by *Fusarium*, *Myrothecium*, *Phomopsis*, *Stachybotrys*, *Trichoderma* and *Trichothecium* spp. Advantage of being relatively cheaper production, extreme toxicity and stability at room temperature for years drives their usage in biological warfare. Previous toxicity studies have demonstrated that these amphipathic molecules can easily cross cell membrane and absorb via integumentary, gastrointestinal systems, skin and rapidly proliferating tissues. Inside the eukaryotic host tissues, trichothecenes interfere host cell protein synthesis by inhibiting the peptide bond formation at peptidyl transferase of 60S ribosomal subunit. Prolonged exposure of trichothecene may cause severe health issues including nausea, vomiting, skin dermatitis, diarrhea and hemorrhagic lesions [97]. Trichothecene also exerts adverse effects on the host immune system. Common trichothecenes include HT-2 toxin, T-2 toxin, nivalenol, diacetoxyscirpenol and deoxynivalenol (also known as "DON" or vomitoxin). The "DON" contamination is common in grains, including barley, oats, rye and wheat [98].

5.11.9 Zearalenone

Zearalenone is a class of nonsteroidal estrogen or phytoestrogen mycotoxins produced by genus *Fusarium* including *F. crookwellense*, *F. culmorum*, *F. cerealis*, *F. equiseti*, *F. graminearum* (Gibberellazeae) and *F. semitectum*. It is synthesized by *Fusarium* species by following polyketide pathway. Zearalenone is among the common cereal crops contaminant (barley, oats, wheat, sorghum, millet and rice) worldwide and implicated adverse effects on reproductive health of farm animals.

Prolonged consumption of zearalenone-contaminated moldy corn and cereal causes hyperestrogenic syndromes in humans. After oral administration, zearalenone rapidly absorbed and underwent two major biotransformation pathways resulting in the formation of α-ZEA and β-ZEA and subsequently conjugated with glucuronic acid. Consumed ZEA and their metabolites cause acute, sub-acute and chronic toxicities to the reproductive system and host immunity complexes. The common symptoms associated with hyperestrogenism syndrome include enlargement of the uterus, nipples, vaginal prolapse and also infertility [99, 100]. The FAO/WHO Joint Expert Committee on Food Additives has established that a maximum tolerable dose of zearalenone is 0.5 µg/kg body weight.

5.12 Antifungal agents

Antifungal medications are the spectrum of pharmaceutical fungistatic or fungicidal components applied for the prevention and cure of superficial, subcutaneous and deep mycosis. The antifungal agents are unlike antibacterial agents, and a relatively very limited progress has been identified in the discovery of novel antifungal drugs. In 1980, all antifungal drugs were broadly categorized into four major classes: allylamines, azoles, morpholines and polyenes, and yet the only new drug introduced for the treatment of systemic fungal infections was oral ketoconazole. It would be more than 10 years before either fluconazole or itraconazole became available for the treatment of systemic mycoses.

5.12.1 Allylamines

Allylamines are synthetic antifungal components effective against a range of pathogenic fungi, particularly for the topical and oral treatment of dermatophytes. They are characterized by interfering ergosterol biosynthesis and blocking squalene metabolism which leads to membrane disruption and cell death in susceptible fungi. Among different identified allylamines, naftifine and terbinafine are drugs of choice and licensed for the clinical application. Naftifine is primarily fungicidal against yeasts as well as against dermatophytes. It is only available as a topical preparation. At higher concentrations, naftifine was found effective against *C. albicans* in several *in vitro* experimental models and may be clinically applicable against this organism. Whereas terbinafine (Lamisil; Novartis Pharmaceuticals) is an allylamine drug broadly used in the treatment of superficial fungal infections caused by dermatophytes. It also has some protective efficacy against the infection of *Aspergillus* species. *Candida* species, *Histoplasma capsulatum*, *Blastomyces dermatitidis*, *Paracoccidioides brasiliensis*, *Sporothrix schenckii* and *Talaromyces* (*Penicillium*) *marneffei*.

The orally administered terbinafine is rapidly absorbed and disseminates to the distant parts of the body. Therefore, it is preferred for dermatophyte infections of the nails and skin. Terbinafine usually causes moderate side effects, including nausea, abdominal cramp, diarrhea and skin rashes. Even prolonged exposure of the terbinafine rarely develops any antifungal resistance in the host.

5.12.2 Azole antifungal drugs

Azole represents a large group of synthetic antifungal components effective against topical treatment of dermatophyte infections and superficial forms of candidiasis. The members of azole anti-fungal agents are broadly categorized into imidazoles (i.e., clotrimazole, ketaconazole and miconazole) and triazoles (e.g., itraconazole and fluconazole) [101]. Apart from ketoconazole, application of imidazoles is limited to the treatment of topical mycoses, whereas triazoles have a broad range of applications in the treatment of superficial, subcutaneous and deep tissue mycosis. Triazoles have demonstrated heightened affinity for fungal than the mammalian P-450 enzymes, hence, confirm least side effects to the host cells. The molecular analysis of azole action in infected host cells displayed inhibition of fungal cytochrome P-450-dependent enzyme, lanosterol 14 α-demethylase, and leads to the accumulation of different methylated sterols (toxic) and exhaustion of ergosterol [101, 102]. This interference in the ergosterol biosynthesis resulted in the disruption of fungal cell membrane structure and also functions. The activity is basically fungistatic, although itraconazole and voriconazole can show fungicidal effects against *Aspergillus* and also against some other molds at specific concentrations with recommended dosages.

5.12.3 Griseofulvin antifungal drug

Griseofulvin has been used as an antifungal antibiotic for the treatment of mycotic diseases of humans and veterinary animals. It is the first antifungal agent tested and claimed to have a selective inhibitory activity against fungal infection in humans. Secondary metabolites obtained from fungus *Penicillium griseofulvum* are used to block the microtubules assembly in susceptible fungal species. The griseofulvin is specifically reactive against pathogenic dermatophytes causing ringworm infections of hair, nails and skin [103, 104]. Oral intake of the component in single and multiple dose form demonstrated effective absorption and dissemination of drug to the distal end of the body in host. It has also proven efficient toward the successful treatment of ringworm of scalp.

5.12.4 Polyene antifungal drugs

Polyenes are one of the most effective antifungal agents naturally obtained from different species of *Streptomyces* species. It consists of a closed macrolide lactone ring with lipophilic side chain contriving conjugated double bonds on one side and hydroxyl groups on the opposite side. This amphipathic nature of polyenes is considered important due to its mechanism of action. Complex amphipathic structure of polyene helps to interact with sterols in the cell membrane (ergosterol in fungi) and adversely affects the membrane barrier, leakage of cellular component, disrupted metabolism and death. It may also stimulate cellular oxidative stress via cascade of reaction and causing extensive damage in the susceptible cells. Amphotericin B, mepartricin, nystatin and natamycin are clinically approved polyenes used in their derivative forms for the treatment of systemic, vaginal, oral and ocular fungal infections. Polyenes are found to be effective against a wide range of pathogenic fungi, including *Aspergillus* species, *Candida* species, mucoraceous molds, dimorphic fungi (*B. dermatitidis*, *Coccidioides* species, *P. brasiliensis* and *H. capsulatum*) and many dematiaceous fungi [105, 106]. Polyenes are still considered as a prominent choice of treatment for several serious fungal infections, which include coccidioidomycosis, blastomycosis, sporotrichosis, histoplasmosis, cryptococcosis and also mucormycosis. However, due to the introduction of voriconazole and echinocandins, amphotericin B is not regarded as the main preferred drug for many cases of aspergillosis and/ or candidiasis [106].

5.12.5 Echinocandins

Echinocandin is the newest antifungal class, introduced in 2001. It is basically semisynthetic in nature that functions by disrupting fungal cell wall biosynthesis. Supplementation of echinocandins interferes in the formation of essential cell wall polysaccharide 1,3-β-D-glucan by noncompetitive inhibition of enzyme 1,3-β-D-glucan synthase and eventually leads to fungal cell death. With the sharing mechanism of action like penicillin against bacterial infection, echinocandins are also called as "penicillin of antifungal." These drugs are prescribed only for the intravenous administration due to low oral absorption and high molecular weight. Anidulafungin, caspofungin and micafungin are echinocandin derivatives utilized in the treatment of several severe fungal infections, especially in the treatment of candidiasis [107, 108]. The clinically available echinocandins are also active against *Aspergillus* species, even to the amphotericin B-resistant strains. However, it is found ineffective against fungi species devoid of 1,3-β-D-glucan in their cell wall, including *C. neoformans* and *Trichosporon* species, as well as *Fusarium* species and the mucoraceous molds. Some findings demonstrated *in vitro* supplementation of micafungin, which inhibits mycelial forms of dimorphic fungi but is ineffective in *in vivo* condition in the tissue forms.

The clinical application of echinocandins via intravenous route is well adapted, and their potential use is related with very few significant unfavorable effects [107]. The common side effect is gastrointestinal (in nature), but in very small percentage of patients (1–5%). Administration of anidulafungin and micafungin may cause infusion-related pain and phlebitis, but these are less common in case of caspofungin. Transient increase in liver enzymes has also been reported in a few cases.

5.12.6 Flucytosine

Flucytosine is a semisynthetic fluorinated pyrimidine (5-fluorocytosine) converted to fluorouracil by cytosine deaminase. It impairs cellular nucleic acid and protein synthesis in selective pathogenic yeasts. The spectrum of antifungal activity includes majority of *Candida* species, however retains very rare clinical application against candidiasis. Rapid development of resistant *Candida* species could be the major obstacle for its clinical application. Patients with the infection of *Cryptococcus* spp. are effectively cured by flucytosine, where it is prescribed along with amphotericin B. It is also recommended as the first-line therapy for the treatment of cryptococcal meningitis. Oral administration along with amphotericin B during the induction period is found protective. Flucytosine does not have any protective effect against infection caused by dimorphic fungi or filamentous fungal pathogens and demonstrated least hindrance with CYP450 enzymes of the host cell. Flucytosine accumulates ubiquitously in almost all host compartments, however, predominantly stored in cerebrospinal fluid and vitreal fluid. Heightened accumulation of drug develops flucytosine-associated toxicities in the host [109]. It particularly causes bone marrow suppression and liver toxicity. Both peak drug levels and cell counts should be monitored during therapy. Dose reductions are often needed. Oral administration may also be associated with gastrointestinal upset and rash. It is metabolically inert and primarily excreted renally; hence, it exhibits excellent antifungal activity in the urinogenital tract.

References

[1]   Hawksworth DL, Pandora's mycological box: Molecular sequences vs. morphology in understanding fungal relationships and biodiversity, Rev Iberoam Micol, 2006, 23, 127–33.

[2]   Levinson W, Review of Medical Microbiology and Immunology, New York, United States, The McGraw-Hill Medical, 2008.

[3]   Hawksworth DL, A new dawn for the naming of fungi: Impacts of decisions made in Melbourne in July 2011 on the future publication and regulation of fungal names, IMA Fungus, 2011, 2, 155–62.

[4]   Salkin IF, Gordon MA, Polymorphism of *Malassezia furfur*, Can J Microbiol, 1977, 23, 471–75.

[5]   Schwartz RA, Superficial fungal infections, The Lancet, 2004, 364, 1173–82.

[6]   Ingham E, Cunningham AC, Malassezia furfur, J Med Vet Mycol, 1993, 31, 265–88.

[7]   Eggimann P, Garbino J, Pittet D, Epidemiology of Candida species infections in critically ill non-immunosuppressed patients, Lancet Infect Dis, 2003, 3, 685–702.

[8]   Lakshmipathy DT, Kannabiran K, Review on dermatomycosis: Pathogenesis and treatment, Natural Science, 2010, 2, 726–31.

[9]   Borman AM, Summerbell RC, Trichophyton, Microsporum, Epidermophyton, and agents of superficial mycoses, Jorgensen JH, Pfaller MA, Carroll KC, Funke G, Landry ML, Richter SS, Warnock DW, ed, Manual of Clinical Microbiology, Vol. 2, Washington DC, ASM Press, 2015, 2128–52.

[10] Dias MFRG, Quaresma-Santos MVP, Bernardes-Filho F, Amorim AGDF, Schechtman RC, Azulay DR, Update on therapy for superficial mycoses: Review article part I, An Bras Dermatol, 2013, 88, 764–74.

[11] Bustamante B, Campos PE, Endemic sporotrichosis, Curr Opin Infect Dis, 2001, 14, 145–49.

[12] Smith MB, McGinnis MR, Subcutaneous Fungal Infections (Chromoblastomycosis, Mycetoma, and Lobomycosis), Hospenthal DR, Rinaldi MG, eds., Diagnosis and Treatment of Human Mycoses, Humana Press, Springer Nature Switzerland AG, 2008, 393–428.

[13] Naggie S, Perfect JR, Molds: Hyalohyphomycosis, phaeohyphomycosis, and zygomycosis, Clin Chest Med, 2009, 30, 337–53.

[14] Kajiwara H, Saito M, Ohga S, Uenotsuchi T, Yoshida SI, Impaired host defense against *Sporothrixschenckii*in mice with chronic granulomatous disease, Infect Immun, 2004, 72, 5073–79.

[15] Almeida OP, Jorge JJ, Scully C, Paracoccidioidomycosis of the mouth: An emerging deep mycosis, Crit Rev Oral Biol Med, 2003, 14, 268–74.

[16] Bonifaz A, Vázquez-González D, Perusquía-Ortiz AM, Endemic systemic mycoses: Coccidioidomycosis, histoplasmosis, paracoccidioidomycosis and blastomycosis, Journal Der DeutschenDermatologischen Gesellschaft, 2011, 9, 705–15.

[17] Ascioglu S, Rex JH, De Pauw B, Bennett JE, Bille J, Crokaert F, et al, Defining opportunistic invasive fungal infections in immunocompromised patients with cancer and hematopoietic stem cell transplants: An international consensus, Clin Infect Dis, 2002, 34, 7–14.

[18] Naiker S, Odhav B, Mycotic keratitis: Profile of *Fusarium* species and their mycotoxins, Mycoses, 2004, 47, 50–56.

[19] Supparatpinyo K, Khamwan C, Baosoung V, Sirisanthana T, Nelson KE, Disseminated *Penicillium marneffei* infection in southeast Asia, The Lancet, 1994, 344, 110–13.

[20] Chakrabarti A, Shivaprakash MR, Microbiology of systemic fungal infections, J Postgrad Med, 2005, 51, 16–20.

[21] Brown GD, Denning DW, Gow NA, Levitz SM, Netea MG, White TC, Hidden killers: Human fungal infections, Sci Transl Med, 2012, 4, 165rv13.

[22] Ribes JA, Vanover-Sams CL, Baker DJ, Zygomycetes in human disease, Clin Microbiol Rev, 2000, 13, 236–301.

[23] Chayakulkeeree M, Ghannoum MA, Perfect JR, Zygomycosis: The re-emerging fungal infection, Eur J Clin Microbiol Infect Dis, 2006, 25, 215–29.

[24] Park BJ, Wannemuehler KA, Marston BJ, Govender N, Pappas PG, Chiller TM, Estimation of the current global burden of cryptococcal meningitis among persons living with HIV/AIDS, Aids, 2009, 23, 525–30.

[25] Kaufman L, Immunohistologic diagnosis of systemic mycoses: An update, Eur J Epidemiol, 1992, 8, 377–82.

[26] Gonçalves AB, Santos IM, Paterson RRM, Lima N, FISH and Calcofluor staining techniques to detect in situ filamentous fungal biofilms in water, Rev Iberoam Micol, 2006, 23, 194–98.

[27] Andrews CP, Weiner MH, Immunodiagnosis of invasive pulmonary aspergillosis in rabbits: Fungal antigen detected by radioimmunoassay in bronchoalveolar lavage fluid, Am Rev Respir Dis, 1981, 124, 60–64.

[28] Wheat J, Wheat H, Connolly P, Kleiman M, Supparatpinyo K, Nelson K, et al., Cross-reactivity in *Histoplasma capsulatum* variety capsulatum antigen assays of urine samples from patients with endemic mycoses, Clin Infect Dis, 1997, 24, 1169–71.

[29] Yeo SF, Wong B, Current status of nonculture methods for diagnosis of invasive fungal infections, Clin Microbiol Rev, 2002, 15, 465–84.

[30] Franklyn KM, Warmington JR, Ott AK, Ashman RB, An immunodominant antigen of *Candida albicans* shows homology to the enzyme enolase, Immunol Cell Biol, 1990, 68, 173–78.

[31] Hamilton AJ, Serodiagnosis of histoplasmosis, paracoccidioidomycosis and penicilliosis marneffei; current status and future trends, Med Mycol, 1998, 36, 351–64.

[32] Mitsutake K, Miyazaki T, Tashiro T, Yamamoto Y, Kakeya H, Otsubo T, et al., Enolase antigen, mannan antigen, Cand-Tec antigen, and beta-glucan in patients with candidemia, J Clin Microbiol, 1996, 34, 1918–21.

[33] Bloomfield N, Gordon MA, Elmendorf DF Jr, Detection of *Cryptococcus neoformans* antigen in body fluids by latex particle agglutination, Proc Soc Exp Biol Med, 1963, 114, 64–67.

[34] Khot PD, Fredricks DN, PCR-based diagnosis of human fungal infections, Expert Rev Anti Infect Ther, 2009, 7, 1201–21.

[35] Tsui CK, Woodhall J, Chen W, Andrélévesque C, Lau A, Schoen CD, De Hoog SG, Molecular techniques for pathogen identification and fungus detection in the environment, IMA Fungus, 2011, 2, 177–89.

[36] Iwen PC, Hinrichs SH, Rupp ME, Utilization of the internal transcribed spacer regions as molecular targets to detect and identify human fungal pathogens, Med Mycol, 2002, 40, 87–109.

[37] Einsele H, Hebart H, Roller G, Löffler J, Rothenhofer I, et al., Detection and identification of fungal pathogens in blood by using molecular probes, J Clin Microbiol, 1997, 35, 1353–60.

[38] Radford SA, Johnson EM, Leeming JP, Millar MR, Cornish JM, Foot AB, Warnock DW, Molecular epidemiological study of *Aspergillus fumigatus* in a bone marrow transplantation unit by PCR amplification of ribosomal intergenic spacer sequences, J Clin Microbiol, 1998, 36, 1294–99.

[39] Santos C, Paterson RRM, Venâncio A, Lima N, Filamentous fungal characterizations by matrix-assisted laser desorption/ionization time-of-flight mass spectrometry, J Appl Microbiol, 2010, 108, 375–85.

[40] Chalupová J, Raus M, Sedlářová M, Šebela M, 2014, Identification of fungal microorganisms by MALDI-TOF mass spectrometry, Biotechnol Adv, 2014, 32, 230–41.

[41] Untereiner WA, Scott JA, Naveau FA, Sigler L, Bachewich J, Angus A, The Ajellomycetaceae, a new family of vertebrate-associated Onygenales, Mycologia, 2004, 96, 812–21.

[42] Mandel MA, Barker BM, Kroken S, Rounsley SD, Orbach MJ, Genomic and population analyses of the mating type loci in *Coccidioides* species reveal evidence for sexual reproduction and gene acquisition, Eukaryot Cell, 2007, 6, 1189–99.

[43] Teixeira MM, Theodoro RC, Nino-Vega G, Bagagli E, Felipe MS, *Paracoccidioides* species complex: Ecology, phylogeny, sexual reproduction, and virulence, PLoS Pathog, 2014, 10, e1004397.

[44] Retallack DM, Woods JP, Molecular epidemiology, pathogenesis, and genetics of the dimorphic fungus *Histoplasma capsulatum*, Microbes Infect, 1999, 1, 817–25.

[45] Kauffman CA, Histoplasmosis, Kauffman CA, Pappas PG, Sobel JD, Dismukes WE, ed, Essentials of Clinical Mycology, New York, USA, Springer-Verlag, 2011, 321–35.

[46] Lenhart SW, Schafer MP, Singal M, Hajjeh RA, Histoplasmosis: Protecting Workers at Risk, 2004, DHHS (NIOSH) Publication No. 2005-109, Washington, DC, National Institute for Occupational Safety and Health.

[47] Smith JA, Kauffman CA, Blastomycosis, Proc Am Thorac Soc, 2010, 7, 173–80.

[48] Reed KD, Meece JK, Archer JR, Peterson AT, Ecologic niche modeling of *Blastomyces dermatitidis* in Wisconsin, PloS One, 2008, 3, e2034.

[49] Tripathy U, Yung GL, Kriett JM, Thistlethwaite PA, Kapelanski DP, Jamieson SW, Donor transfer of pulmonary coccidioidomycosis in lung transplantation, Ann Thorac Surg, 2002, 73, 306–08.

[50] Thompson GR, Pulmonary coccidioidomycosis, Semin Respir Crit Care Med, 2011, 32, 754–63.

[51] Anstead GM, Sutton DA, Graybill JR, Adiaspiromycosis causing respiratory failure and a review of human infections due to Emmonsia and Chrysosporium spp, J Clin Microbiol, 2012, 50, 1346–54.

[52] El-Zammar OA, Katzenstein AL, Pathological diagnosis of granulomatous lung disease: A review, Histopathol, 2007, 50, 289–310.

[53] Echavarria E, Cano EL, Restrepo A, Disseminated adiaspiromycosis in a patient with AIDS, J Med Vet Mycol, 1993, 31, 91–97.

[54] Sigler L, Ajellomycescrescens sp. nov., taxonomy of *Emmonsia* spp., and relatedness with *Blastomyces dermatitidis* (teleomorph *Ajellomyces dermatitidis*), J Med Vet Mycol, 1996, 34, 303–14.

[55] Peterson SW, Sigler L, Molecular genetic variation in *Emmonsia crescens* and *Emmonsia parva*, etiologic agents of Adiaspiromycosis, and their phylogenetic relationship to *Blastomyces dermatitidis* (Ajellomyces dermatitidis) and other systemic fungal pathogens, J Clin Microbiol, 1998, 36, 2918–25.

[56] Fisher MC, Koenig GL, White TJ, Taylor JW, Molecular and phenotypic description of *Coccidioides posadasii* sp. nov., previously recognized as the non-California population of *Coccidioides immitis*, Mycologia, 2002, 94, 73–84.

[57] Galgiani JN, Ampel NM, Blair JE, Catanzaro A, Johnson RH, Stevens DA, et al., Coccidioidomycosis, Clin Infect Dis, 2005, 41, 1217–23.

[58] Hector RF, Laniado-Laborin R, Coccidioidomycosis – a fungal disease of the Americas, PLoS Med, 2005, 2, e2.

[59] Galgiani JN, Ampel NM, Catanzaro A, Johnson RH, Stevens DA, Williams PL, Practice guidelines for the treatment of coccidioidomycosis, Clin Infect Dis, 2000, 30, 658–61.

[60] Awasthi S, Cox RA, Transfection of murine dendritic cell line (JAWS II) by a nonviral transfection reagent, Biotechniques, 2003, 35, 600–05.

[61] Gugnani HC, Entomophthoromycosis due to Conidiobolus, Eur J Epidemiol, 1992, 8, 391–96.

[62] Pfaller MA, Diekema DJ, Unusual fungal and pseudofungal infections of humans, J Clin Microbiol, 2005, 43, 1495–504.

[63] Vismer HF, DeBeer HA, Dreyer L, Subcutaneous phycomycosis caused by *Basidiobolus* Haptosporus (Drechsler 1947), S Afr Med J, 1980, 58, 644–47.

[64] Burns RA, Roy JS, Woods C, Padhye AA, Warnock DW, Report of the first human case of lobomycosis in the United States, J Clin Microbiol, 2000, 38, 1283–85.

[65] Rodríguez G, Barrera GP, The asteroid body of lobomycosis, Mycopathologia, 1996, 136, 71–74.

[66] Migaki G, Valerio MG, Irvine B, Garner FM, Lobo's disease in an Atlantic bottle-nosed dolphin, J Am Vet Med Assoc, 1971, 159, 578–82.

[67] De Vries GA, Laarman JJ, A case of Lobo's disease in the dolphin Sotalia guianensis, Aquat Mamm, 1973, 1, 26–33.

[68] Restrepo A, Treatment of tropical mycoses, J Am Acad Dermatol, 1994, 31, S91–S102.

[69] Taborda V, Taborda PR, Mcginnis MR, Constitutive melanin in the cell wall of the etiologic agent of Lobo's disease, Revista Do Instituto De Medicina Tropical De São Paulo, 1999, 41, 9–12.

[70] Ali A, Flieder D, Guiter G, Hoda SA, Rhinosporidiosis: An unusual affliction, Arch Pathol Lab Med, 2001, 125, 1392–93.

[71] Loh KS, Chong SM, Pang YT, Soh K, Rhinosporidiosis: Differential diagnosis of a large nasal mass, Otolaryngology-Head and Neck Surgery, 2001, 124, 121–22.

[72] Das S, Kashyap B, Barua M, Gupta N, Saha R, Vaid L, Banka A, Nasal rhinosporidiosis in humans: New interpretations and a review of the literature of this enigmatic disease, Med Mycol, 2011, 49, 311–15.

[73] Arseculeratne SN, Recent advances in rhinosporidiosis and *Rhinosporidium seeberi*, Indian J Med Microbio, 2002, 20, 119–31.

[74] Arseculeratne SN, Bact D, Phil D, Chemotherapy of rhinosporidiosis: A review, J Infect Dis Antimicrob Agents, 2009, 26, 21–27.

[75] Yu J, Current understanding on aflatoxin biosynthesis and future perspective in reducing aflatoxin contamination, Toxins, 2012, 4, 1024–57.

[76] Wogan GN, Chemical nature and biological effects of the aflatoxins, Bacteriol Rev, 1966, 30, 460.

[77] Gnonlonfin GJB, Hell K, Adjovi Y, Fandohan P, Koudande DO, Mensah GA, et al., A review on aflatoxin contamination and its implications in the developing world: A sub-Saharan African perspective, Crit Rev Food Sci Nutr, 2013, 53, 349–65.

[78] Sun G, Wang S, Hu X, Su J, Zhang Y, Xie Y, et al., Co-contamination of aflatoxin B1 and fumonisin B1 in food and human dietary exposure in three areas of China, Food Addit Contam, 2011, 28, 461–70.

[79] Holzapfel CW, The isolation and structure of cyclopiazonic acid, a toxic metabolite of *Penicillium cyclopium* Westling, Tetrahedron, 1968, 24, 2101–19.

[80] Purchase IFH, The acute toxicity of the mycotoxin cyclopiazonic acid to rats, Toxicol Appl Pharmacol, 1971, 18, 114–23.

[81] Dorner JW, Cole RJ, Lomax LG, Gosser HS, Diener UL, Cyclopiazonic acid production by Aspergillus flavus and its effects on broiler chickens, Appl Environ Microbiol, 1983, 46, 698–703.

[82] Nuehring LP, Rowland GN, Harrison LR, Cole RJ, Dorner JW, Cyclopiazonic acid mycotoxicosis in the dog, American J Veter Res, 1985, 46, 1670–76.

[83] Lomax LG, Cole RJ, Dorner JW, The toxicity of cyclopiazonic acid in weaned pigs, Vet Pathol, 1984, 21, 418–24.

[84] Antony M, Shukla Y, Janardhanan KK, Potential risk of acute hepatotoxicity of kodo poisoning due to exposure to cyclopiazonic acid, J Ethnopharmacol, 2003, 87, 211–14.

[85] Bennett JW, Bentley R, Pride and prejudice: The story of ergot, Perspect Biol Med, 1999, 42, 333–55.

[86] Panaccione DG, Coyle CM, Abundant respirable ergot alkaloids from the common airborne fungus *Aspergillus fumigatus*, Appl Environ Microbiol, 2005, 71, 3106–11.

[87] Marasas WF, Kellerman TS, Pienaar JG, Naude TW, Leukoencephalomalacia: A mycotoxicosis of Equidae caused by *Fusarium moniliforme* Sheldon, Onderstepoort J Vet Res, 1976, 43, 113–22.

[88] Harrison LR, Colvin BM, Greene JT, Newman LE, Cole JR Jr, Pulmonary edema and hydrothorax in swine produced by fumonisin B1, a toxic metabolite of *Fusarium moniliforme*, J Vet Diagn Invest, 1990, 2, 217–21.

[89] Marasas WF, Kriek NP, Fincham JE, Van Rensburg SJ, Primary liver cancer and oesophageal basal cell hyperplasia in rats caused by *Fusarium moniliforme*, Int J Cancer, 1984, 34, 383–87.

[90] Smith JE, Moss MO, *Mycotoxins*. Formation, Analysis and Significance, 148 S., 54 Abb., 52 Tab, Chichester-New York-Brisbane-Toronto-Singapore, John Wiley & Sons, 1985.

[91] Marquardt RR, Frohlich AA, A review of recent advances in understanding ochratoxicosis, J Anim Sci, 1992, 70, 3968–88.

[92] Sava V, Reunova O, Velasquez A, Sanchez-Ramos J, Can low level exposure to ochratoxin-A cause parkinsonism?, J Neurol Sci, 2006, 249, 68–75.

[93] Norstadt FA, McCalla TM, Patulin production by *Penicillium urticae* Bainier in batch culture, Appl Microbiol, 1969, 17, 193–96.

[94] Van Egmond HP, Schothorst R, Jonker M, Regulations relating to mycotoxins in food Perspectives in a global and European context, Anal Bioanal Chem, 2007, 389, 147–57.

[95] Puel O, Galtier P, Oswald IP, Biosynthesis and toxicological effects of patulin, Toxins, 2010, 2, 613–31.

[96] Trucksess MW, Tang Y, Solid phase extraction method for patulin in apple juice and unfiltered apple juice, J AOAC Int, 1999, 82, 1109–13.

[97] Takayama H, Shimada N, Mikami O, Murata H, Suppressive effect of deoxynivalenol, a *Fusarium* mycotoxin, on bovine and porcine neutrophil chemiluminescence: An *in vitro* study, J Vet Med Sci, 2005, 67, 531–33.

[98] Hinkley SF, Mazzola EP, Fettinger JC, Lam YF, Jarvis BB, Atranones A-G, from the toxigenic mold *Stachybotrys chartarum*, Phytochemistry, 2000, 55, 663–73.

[99] El-Nezami H, Polychronaki N, Salminen S, Mykkänen H, Binding rather than metabolism may explain the interaction of two food-grade *Lactobacillus* strains with zearalenone and its derivative α´-zearalenol, Appl Environ Microbiol, 2002, 68, 3545–49.

[100] Dänicke S, Winkler J, Invited review: Diagnosis of zearalenone (ZEN) exposure of farm animals and transfer of its residues into edible tissues (carry over), Food Chem Toxicol, 2015, 84, 225–49.

[101] Sanglard D, Odds FC, Resistance of *Candida* species to antifungal agents: Molecular mechanisms and clinical consequences, Lancet Infect Dis, 2002, 2, 73–85.

[102] Ghannoum MA, Kuhn DM, Voriconazole – Better chances for patients with invasive mycoses, European J Med Res, 2002, 7, 242–56.

[103] Develoux M, Griseofulvin, Annales De Dermatologie Et De Vénéréologie, 2001, 128, 1317–25.

[104] Woyke T, Roberson RW, Pettit GR, Winkelmann G, Pettit RK, Effect of auristatin PHE on microtubule integrity and nuclear localization in *Cryptococcus neoformans*, Antimicrob Agents Chemother, 2002, 46, 3802–08.

[105] Anonymous, Liposomal nystatin – Nyotranw – Antifungal, Drugs Future, 2001, 26, 810.

[106] Dupont B, Overview of the lipid formulations of amphotericin B, J Antimicrob Chemother, 2002, 49, 31–36.

[107] Bossche HV, Echinocandins-an update, Expert Opin Ther Pat, 2002, 12, 151–67.

[108] González GM, Tijerina R, Najvar LK, Bocanegra R, Rinaldi M, Loebenberg D, et al., *In vitro* and *in vivo* activities of posaconazole against *Coccidioides immitis*, Antimicrob Agent Chemother, 2002, 46, 1352–56.

[109] Pfaller MA, Messer SA, Boyken L, Huynh H, Hollis RJ, Diekema DJ, In vitro activities of 5-fluorocytosine against 8,803 clinical isolates of *Candida* spp.: Global assessment of primary resistance using National Committee for Clinical Laboratory Standards susceptibility testing methods, Antimicrob Agent Chemother, 2002, 46, 3518–21.

Saloni Singh
Chapter 6
Microbial assay techniques

Abstract: Microbiology deals with the study of organisms that cannot be visualized by the naked eye due to their very small size. It is very important to visualize, observe, diagnose and analyze the microbes due to their involvement in various diseases. The biochemical tests, staining procedures and critical analysis of microbial characteristics are still very important in microbiology lab. The chapter contains various microbiological stains, procedures of staining, observation of microbes and biochemical assays of microbes. This chapter will be useful for students, academicians and researchers involved in routine assays related to microbes.

Keywords: smears, stains, biochemical tests, microscopy, hyphae, parasites

6.1 Microbiology

Microbiology is the branch of science which involves the study of organisms which cannot be visualized by the naked eye due to their very small size. These microorganisms are taxonomically distributed as bacteria, fungi, algae, protozoa and virus. Almost all microbes are of economic importance due to either their applications in various fields or their potential to cause disease in other organisms.

6.1.1 Microbiological stains

The existence of microorganisms was unknown until the invention of optical instrument microscope. The microorganisms can be visualized under microscope by application of different stains after following a specific staining procedure. The stain and its process vary according to the group of microorganism. The microorganism should be fixed and then stained to enhance visibility, get accurate morphological features and for preservation so as to study in future also. The basic steps before proceeding to the stain specific procedures involves smear preparation and fixation of microorganisms.

Saloni Singh, School of Life Sciences, Jaipur National University, Jaipur 302025, Rajasthan, India, e-mail: singhsaloni1990@gmail.com

https://doi.org/10.1515/9783110517736-006

6.1.1.1 Smear preparation

A smear is an even spread of microbial suspension. It is made by dragging a drop of suspension across clean slide from one corner end to another by using another clean and completely dried slide by making an angle of 45° [1]. It is then preceded for fixation (Figure 6.1).

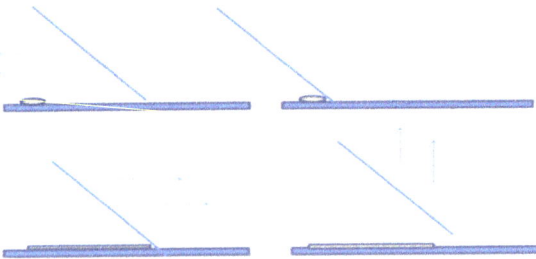

Figure 6.1: Smear preparation.

6.1.1.2 Fixation

Although the living microorganisms can be visualized directly under light microscope but fixation over the glass slide is preferred because fixation over a glass surface improves staining results. Moreover, internal and external structures of microbes can be preserved by fixation [2]. Fixation inactivates the enzymes which may affect the cell morphology. Fixation is of two types, that is, heat fixation and chemical fixation.

6.1.1.2.1 Heat fixation
The glass surface/slide having microbial smear is gently exposed to flame of Bunsen burner in a to and fro manner and the temperature of slide is kept low to avoid charring of microbial smear. It helps in preserving overall morphology.

6.1.1.2.2 Chemical fixation
It protects fine cellular substructure. It penetrates within cell and thus reacts with protein and lipid making them inactive and immobilized. Some commonly used fixatives are ethanol, formaldehyde, glutaraldehyde and acetic acid.

6.1.2 Stains

Dyes are organic compounds that have affinity for cellular material and increases color variations as the metabolites of cell react with dye forming colored complex.

Most of stains/dyes consist of three functional groups (i.e., chromophore, auxochrome and chromogen) and generally formed of colorless organic solvent, chromophore and auxochrome.

1. Chromophore: The ionic component of the dye that imparts color to the cell is known as chromophore. Chromophore belongs to the chemical group which is capable of light absorption in selective range of wavelength and results in the coloration of certain compounds. It gives the characteristic color to the dye. It stains in both positive and negative form.

2. Auxochrome: It imparts a particular color when attached to a chromophore but in absence of chromophore, it fails to produce that color. It intensifies the color and enables the chromogen to form salts and binding to biological molecules.

3. Chromogen: It is an organic compound that can be converted in a pigment or dye. It may be defined as a substance capable of conversion into a pigment or dye having benzene ring and a chromophore.

Dyes have been classified as acidic or basic depending upon the charge over the pigment, that is, positive or negative [3]. Based on the charge, dyes can be anionic, cationic or neutral.

A. Cationic dyes: The commonly used positively charged (or basic) dyes are methylene blue, crystal violet, malachite green and also include safranin. The basic dyes combine with those cellular elements that are acidic in nature. The dyes stain negatively charged structures like cell wall, nucleic acids and some proteins.

B. Anionic dyes: These dyes include negatively charged (acidic dyes) India ink and nigrosin. These dyes interact with positively charged molecules but are repelled by negative charge on surface of the cell. Therefore, background gets colored and cell remains unstained. These stains are generally used to study morphological characters including shape, size and also the various arrangements of bacteria.

C. Neutral dyes: They are a mixture of aqueous solution of acidic and basic dyes resulting in the formation of neutral dye.

6.2 Bacteriology

The study of bacteria is known as bacteriology. Bacteriology includes the identification, biochemical studies, molecular studies and cultivation of bacterial cells. Bacteriology also includes applications of bacteria in various fields including medicine, agriculture sector, pharmaceutical industry and applied biotechnology. Bacteria can be identified and characterized on the basis of staining, microscopy, biochemical tests and specific nucleotide sequencing. Bacteria are prokaryotes which lack membrane-bound nuclei. Their DNA exists in dense cytoplasm region known as nucleoid [4]. Bacterial cells may contain small circular and extra chromosomal DNA (plasmids) that can be shared from one cell to another. It is easy to transform the bacteria with a foreign gene of interest. Most of molecular biology and biotechnology approaches depend on the bacterial genetic modifications. It also makes bacteria a favorite of molecular biology workers and researchers working in area of genetic engineering due to ease of their genetic manipulations [5].

Bacteria lack membrane-bound well-defined organelles similar to mitochondria or chloroplasts present in eukaryotes. Some photosynthetic bacteria (like cyanobacteria) may have tightly packed folds of their outer membrane [6]. Bacterial cells generally have two basic types of shapes, that is, spherical (known as coccus) or the rod-shaped (known as bacillus). These rod-shaped bacterial cells further [7] may show variations in shapes including spiral, (spirillum and spirochetes) comma-shaped (vibrio) or filamentous. The cytoplasm is an aqueous environment having cytoplasmic content like ribosomes and a large number of different proteins and also nucleotide–protein complexes. Some larger structural organizations including pores or inclusions in forms of granules containing storage products may also be present in some species under specific growth conditions [8].

6.3 Staining

6.3.1 Types of staining

Simple staining: Microorganism can be stained by single dye or a combination of dyes [9]. Simple staining is known as monochrome (single dye) staining or positive staining. In this method of staining, bacterial cells are stained using a suitable single stain (Figure 6.2). Commonly used simple stains are crystal violet, methylene blue, malachite green, safranin and basic fuchsin. In simple stain, microbial cells can be stained uniformly.

Differential staining: This process of staining explores the advantage of differences in the properties (physical and chemical) of different groups of bacteria. In this type of staining, two different stains of different colors are used either in combination (simultaneously) or in a successive manner (one followed by other).

Differential staining allows us to differentiate the various kinds of bacterial cells and also different parts of a bacterial cell. Gram's staining process is the most commonly used differential staining. It was initially described by Christian Gram in 1884. The Gram's stain process divides the bacteria broadly into two groups: (i) gram-positive bacteria and (ii) gram-negative bacteria. The microorganisms which are able to hold the primary stain (crystal violet) appear violet or purple. These cells are designated as gram-positive cells. On the other side, the gram-negative cells lose the crystal violet and subsequently adopt the second stain known as safranin (as counterstain). These gram-negative cells appear reddish/pink in color (Figure 6.3).

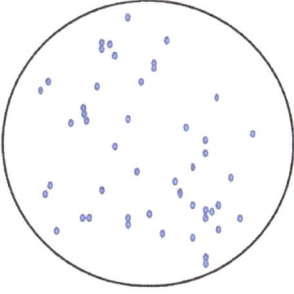

Figure 6.2: Simple or monochrome stained bacterial cells.

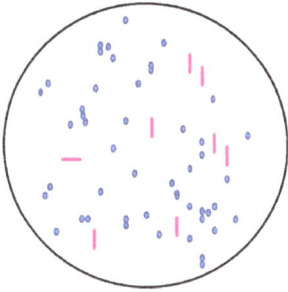

Figure 6.3: Differential staining revealing two categories of cells, that is, gram-positive and gram-negative cells.

Many other special staining procedures have also been introduced to study specific structures such as capsule, endospore, storage granules and flagella.

6.3.2 Differential staining

6.3.2.1 Gram staining

The Gram staining method was developed by Hans Christian Gram (the Danish bacteriologist). In 1884, he developed this differential staining technique [10]. He, for the first time, used this method for staining the *Klebsiella pneumoniae*. This process is a multistep and sequential staining protocol which separates the bacteria into four major groups depending on the cell morphology and structure of cell wall. These four groups are (i) gram-positive (gram +ve) cocci, (ii) gram-negative (gram −ve) cocci, (iii) gram-positive (gram +ve) rods and (iv) gram-negative (gram −ve) rods [10]. This method is based on the application of two dyes, which are crystal violet and safranin [11], in a sequential manner for a specified time. The microbial cells which take first/primary stain were called as gram-positive while those retained second stain were termed as gram negative. Gram-positive cells give violet or purple appearance while gram-negative cells appear reddish or dark pink in color during microscopic examination. The gram stain is very useful in routine cell culture work for assessing the purity and contamination of culture samples. The gram stain is also very important in primary identification of bacteria isolated from mixed population or natural reservoirs [10]. It is still the most widely employed method for visualizing bacteria.

The principle for differential staining in gram stain lies in chemical composition of bacterial cell wall and components used in gram staining. In gram staining, four components are used (in sequence). These stains include primary stain (crystal violet) followed by mordant (iodine), thereafter, decolorizer (ethanol/acetone) and the last counterstain (safranin). Bacterial cell wall is composed of peptidoglycan. The cell wall gives rigidity, firmness and protection to the internal structure of cell. Gram-positive bacteria have comparatively thicker cell wall of peptidoglycan while the gram-negative bacteria have thin cell wall and also have an outer additional membrane based on phospholipids bilayer [12]. The crystal violet stain enters through the cell wall of both types of cells, but the existence of iodine–dye complex remains difficult [13]. In the next step, the use of decolorizer (alcohol or acetone) dissolves the lipids of membranes. The decolorizing mixture results in dehydration of cell wall, and it also serves as a solvent to wash out the crystal violet–iodine complex. In case of gram-negative bacteria, the decolorizer dissolves the outer bilayer membrane made of phospholipids. Due to dissolution of lipids, the dye–iodine complex escapes during the step of decolorization. Due to the loss of outer membrane and the crystal violet–iodine complex, the gram-negative cells are able to take the counter stain (safranin) and therefore, appear pink in color (Figure 6.4). In the case of gram-positive bacteria, the decolorizer dehydrates, shrinks the wall and closes the pores which prevent the leakage of dye–iodine complex; therefore, these cells appear violet or blue in color [14]. There are so many species that may be "gram-variable" with intermediate cell wall structure in addition to the clearly gram positive and gram negative [15].

6.3.2.2 Acid-fast staining

The acid-fast stain is performed on microbial samples to reveal the characteristic of acid-fastness in specific bacteria and also in the cysts of *Cryptosporidium* and *Isospora* [16]. One of the most important clinical applications of acid-fast stain is to detect *Mycobacterium tuberculosis* in provided sputum sample to diagnose tuberculosis [16]. In this staining, basic fuchsin is used as a primary stain. It was used for the first time by Ehrlich in 1882. This method also popularized as the Ziehl–Neelsen method (in the early to mid-1890s). Currently, there are three generally used acid-fast staining methods which include Ziehl–Neelsen (hot) method, Kinyoun (cold) method and the auramine-rhodamine fluorochrome (Truant method).

The Ziehl–Neelsen and the Kinyoun methods are more common due to easy observation of the slides developed in these methods using a standard microscope (bright-field microscopy). The Ziehl–Neelsen method is widely used for routine staining. The fluorochrome-based method is widely used by advanced laboratories

Heat fix the bacterial cells to glass slide

↓

Flood the bacterial smear with crystal violet for 1 minute

↓

Rinse the slide with gentle flow of water; Flood the smear with iodine (mordant) for 1 minute

↓

Rinse off iodine; apply decolorizer to remove excess stain followed by washing with water

↓

Apply counter stain safranin for 30 seconds

↓

Wash with water, dry and observe under microscope

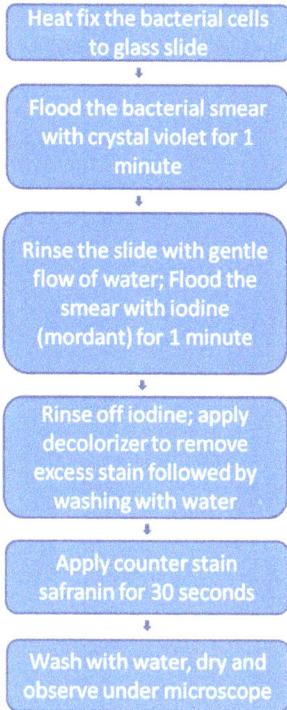

Figure 6.4: Procedure for gram staining.

having fluorescent microscope. It is a heat-based method to impel the primary stain into the waxy cell walls of microbial cells that are otherwise difficult to stain [16]. These staining methods are particularly useful for staining Mycobacterium species and also nontuberculous mycobacteria. The cell wall of mycobacteria contains comparatively higher concentration of lipid therefore, cells are hydrophobic and also impermeable to most routine differential stains such as the gram stain. Furthermore, it has also been found that they are resistant to acid and alcohol and can be described as acid-fast bacilli (AFB) or acid alcohol fast bacilli. This staining process distinguishes AFB from acid non-fast bacteria [17]. In this staining process, microbial cells are primarily stained with a high-reactive primary stain, that is, carbol fuchsin–phenol. The mycolic acid present with cell binds to carbol fuchsin when it penetrates the cell. The phenol–carbol fuchsin stain is then heated to facilitate the penetration of dye to the waxy mycobacterial cell wall. Due to use of heat, this method is called as the "hot staining" method. The stain binds firmly to the mycolic acid present in the mycobacterial cell wall. Thereafter, an acid-based decolorizing solution is applied. It removes the dye from the sample contents (background cells, tissue fibers and any organisms in the smear) except mycobacteria, which retain or hold fast the dye and, therefore, are termed as AFB. The cells having high amount of mycolic acids [18] are tough to decolorize, and therefore, appear pink in color (termed as acid fast organisms). However, non-acid-fast cells are easily decolorized and stained with counter stain, that is, malachite green or methylene blue and therefore, appear green or blue in color, respectively (Figure 6.5). Bacteria described as acid fast will appear red when examining specimens using bright-field microscopy. Non-acid-fast cells and field debris will appear blue.

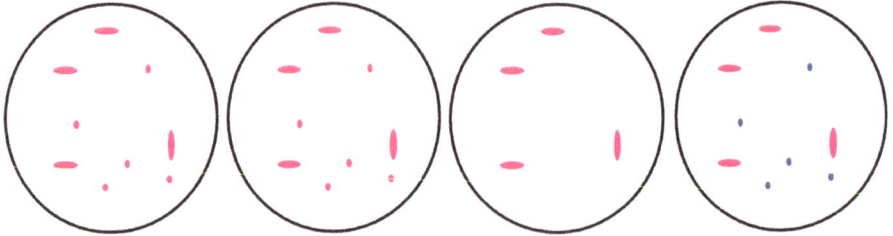

Figure 6.5: Procedure for acid-fast staining involves [23, 24, 16]: (a) application of carbol fuchsin (primary stain); (b) application of heat (mordant); (c) application of acid alcohol (decolorizer) and (d) application of methylene blue.

6.3.3 Special staining procedure

6.3.3.1 Capsule staining (Anthony's method)

Bacterial capsules are generally made of high-molecular-weight polysaccharides and also may contain polypeptides. The capsules are generally associated with virulence and also with bio-film formation [19]. The capsule part of microbial cells does not stain well with routine stains including crystal violet, methylene blue and other simple stains [19]. The bacteria having capsular covering are known as "smooth" while cells lacking the capsule are termed as "rough." Bacterial colonies of the encapsulated bacterial cells are also termed as "smooth" colonies and colonies of non-encapsulated bacteria are known as "rough" colony [20–22].

The capsule stain is used to reveal the bacterial capsule. The capsule staining methods were developed for observation of capsules and these methods provide consistently reliable results [22]. The common capsule staining methods are based on combination of the following [22]; (i) a basic dye; interacts with the –ve ions of the bacterial cell, (ii) a mordant; causes the precipitation of the capsular material [25] and (iii) an acidic stain; to color the background.

The capsule is observed as a clear circle (lightened) between the colored background and the stained cell. In some capsule staining processes, cells are also exposed to antibody against specific capsular antigens to further enlarge the capsule for easier visualization [25].

In Anthony's capsule stain, the crystal violet stain is used as primary stain which interacts with the protein content in the culture broth, and the copper sulfate serves as the mordant. After the staining, the bacterial cells along with background appear stained with the crystal violet and the unstained capsule appears white [22]. In this staining process, a smear of bacterial cells is made at the center of slide. Heat fixation of cells is avoided because heating may cause cell shrinkage and may create a clear zone around the cell [26]. The procedure for this stain involves the following steps [22]. The smear is stained with 1% crystal violet for 2 min. The smear is then rinsed with a 20% solution of copper sulfate. Dry the slide gently in air. Now, the slide can be examined under an oil immersion lens.

6.3.3.2 Capsule staining (India ink method)

India ink is used as negative stain. It is used to identify the presence of capsules. Other stains which may be used are Congo red, nigrosin and eosin. These stains provide color to the background while the capsule can be seen uncolored. The crystal violet stain may be used to stain the cell. India ink and also the other stains (i.e., Congo red, nigrosin), give the background dark appearance. By counterstaining (with dyes like crystal violet or methylene blue), the bacterial cells can be distinctly observed as compared to capsule [27].

6.3.3.3 Endospore staining (the Schaeffer–Fulton method)

Endospores are physiological inactive forms of bacteria that allow it to survive under adverse environmental conditions. The staining process was designed by Alice B. Schaeffer and MacDonald Fulton during the 1930s [28]. This staining procedure is also known by the name Wirtz–Conklin method, referring to two bacteriologists during the 1900s [29]. The Schaeffer–Fulton stain involves staining of endospore by a relatively simple staining procedure. In this staining method, primary stain (malachite green) is forced into the spore through steaming of the bacterial emulsion. Malachite green has a low affinity for cellular material; therefore, vegetative cells may be easily decolorized with water followed by counterstaining with safranin. Spores may be located at center of the cell, at end of the cell or in a position between the end and middle of the cell. The primary stain used in the process is malachite green, while the counter stain is safranin [30]. The procedure involves following steps [29]: Preparation of smear followed by exposure of smear to steam over water bath with help of blotting paper. Next step is to flood the entire slide with malachite green. After 5 min, the slide is removed from the steam, followed by removal of the blotting paper towel. Cool down the slide and rinse with water for approximately 30 s. Apply counterstain safranin for 02 min. The slide is then rinsed again, blotted dry with specific paper and observed under microscope.

6.3.3.4 Storage granules (Albert's staining method)

Some microbial cells possess special internal structures commonly known as granules or inclusions. These are special storage structures generally used during starvation or adverse conditions. The granules are usually made up of poly-metaphosphates and generally known by various names like volutin granules, polar bodies or Babes–Ernst granules. The volutin granules were observed first time in *Spirillum volutans*[31]. Paul Ernst and Victor Babes observed these granules in bacteria stained with methylene blue or toluidine blue. In routine lab practices, the granules are stained generally with Albert's staining method. Albert's staining technique is commonly used to reveal or observe special structure in bacteria. It is used to demonstrate the presence of metachromatic granules present in *Corynebacterium diphtheriae*. The granules appear violet with polychrome methylene blue staining, while the rest of the cell gives blue appearance. Basically, the Albert stain is made up of two stains, that is, toluidine blue and malachite green. The pH of Albert stain is acidic for cytoplasm, whereas it is basic for volutin granules [32]. The Albert's stain gives green color to cytoplasm while granules are stained toluidine blue. Procedure for Albert's stain is as follows [33]:
– Preparation of smear on glass slide
– Air dry and heat fixation of smear
– Treatment of the smear with Albert's stain (3–5 min)
– Excess stain removal
– Treatment of smear with Albert's iodine (1–2 min)
– Washing of the slide with water
– Air dry and observe slide under oil immersion lens

6.3.3.5 Flagella staining (Leifson method)

Bacterial flagella are the appendages which are used for motility [34]. Flagella are one of the most important locomotory organs [35]. Flagella are generally present in rod-shaped bacteria [36]. Flagella are very thin and difficult to be seen by compound light microscopy. The staining method is based on application of a mordant (often tannic acid) to make them observable using an oil immersion objective [34]. Leifson's method [37] is used for staining of flagella. In this method, cell suspension is placed on a clean slide kept in slanting position. The smear is allowed to run down from the inclined slide. The cells are treated with Leifson's flagella stain for 10–15 min. A fine rust-colored precipitate forms throughout the slide. Slide is dried and examined under the oil immersion lens.

6.3.3.6 Hanging drop technique for bacterial motility detection

The motility of bacteria can be evident by observing the presence of flagella [38]. The hanging drop technique enables to study and differentiate the motile bacteria from non-motile bacteria. Motile bacteria have the inherent ability of movement in the surrounding medium [39]. In wet mount method, the motility of bacteria may be hampered as the suspension is pressed between the slide and the cover slip. Therefore, hanging drop test is done for observation of the motility of bacteria. It is helpful in the identification of bacteria [40]. The motile bacteria will show erratic movement in the medium. On the other hand, non-motile will remain static in the medium or may show Brownian movement [41].

6.3.3.6.1 Procedure for the hanging drop technique
- Clean a cavity slide (concave depression slide) thoroughly.
- Apply small amount of petroleum jelly at the corners of coverslip.
- Aseptically transfer two or three loops full of broth culture of motile bacteria in the center of cover slip.
- Invert the cavity slide and place on the cover slip, in a manner that the cavity covers the broth culture of motile bacteria.
- Press the cover slip gently so that petroleum jelly seals the corner of cover slip against the slide.
- The slide is inverted fast, so that the drop hangs into the cavity or concave depression.
- The slide is then clipped to stage (of the microscope).
- The edge of the drop is first focused under low-power objective. Thereafter, switch to high-power objective and observe the motion or motility of bacteria (the motion in a particular direction).
- A drop of immersion oil may be used. Preferably, a phase-contrast or dark-field microscope should be used for better and clear observation [42].
- Also observe shape, size and speed of movement
- This method uses living bacteria therefore, discard the slide accordingly.

6.3.4 Simple staining

It is among the important primary techniques used in routine manner to check the purity of bacterial culture. It is very fast and effective technique to observe the bacterial morphology. This staining technique involves use of single stain, therefore it is known as simple staining. Bacteria cannot be seen with naked eyes because of their shape, size and morphology. Moreover, most bacteria have no color [38]. Therefore, to see bacteria under microscope, it is necessary to stain the bacterial surface with

appropriate color imparting reagent or dye. After staining with specific stain, bacteria can be seen under microscope. The simple staining technique involves use of positive charge carrying stain. The slight negative charge of bacterial surface and cytoplasm attracts the positively charged stain which is further fixed by application of suitable mordant. The procedure for the simple staining of bacterial suspension consists of following steps:
–   Prepare smear of bacterial cells (heat-fixed or air-dried).
–   Cover the smear with basic stain for one minute. During this one minute, do not disturb the slide.
–   Wash off the excess stain with indirect gentle stream of water.
–   Dry the slide with help of absorbent or blotting papers. Do not wipe the slide.
–   Observe the slide under low-power objective followed by high-power objective.
–   Add a drop of immersion oil directly on the smear and switch to oil immersion objective [38].
–   Scan the slide and focus over thin layer of bacteria. Determine the shape, size and arrangement of bacterial cells [38].

6.3.5 Biochemical assay

6.3.5.1 Phenol red broth

It is a routinely used differential test medium for the differentiation of the gram-negative enteric bacteria. This test explores three phenol red broths, that is, lactose, glucose and sucrose. Phenol red is a pH-based indicator which becomes yellow below a pH of 6.8 and pink at upper pH that is pH above 7.4. If the organism utilizes the carbohydrate, an acid is produced; therefore, turn the media yellow in color. If the test organism does not metabolize the carbohydrate but is able to utilize the peptone, then, by-product (ammonia) is produced and this raises the pH of the medium and develops pink color [37].

Gas may be produced as by-product if organism uses the carbohydrate. The air bubbles are trapped in the tube. But, if the test organism is not able to metabolize the carbohydrate, no gas formation and no trapping of air bubbles.

6.3.5.2 Gelatinase test

For detection of gelatinase, the nutrient gelatin is used as differential media. This enzyme hydrolyses gelatin. Gelatin is a semisolid material at a temperature lower than 32 °C. At more than 32 °C, it is liquid in nature. Gelatinase permits the organisms to convert the gelatin into small products including polypeptides, peptides and further into amino acids. These amino acids are able enter in cell by crossing the cell membrane and these are be utilized by the organism. The gelatinase test can be used to differentiate various bacteria [44].

6.3.5.3 Lipase test

Tributyrin agar assesses the ability of an organism to produce lipase. It can hydro-lyze tributyrin oil. Different lipase tests are based on various fat sources. Chiefly, corn oil, peanut oil, olive oil, egg yolk and soybean oil [45] are used. Lipase enables the organisms to break down lipids into smaller fragments. These may be converted into various end products. These end products can be utilized by the cells in various cellular pathways [46].

Tributyrin oil develops an unclear suspension in the agar [47]. The lipase-producing organism breaks down the tributyrin and develops a clear zone surrounding colony or growth.

6.3.5.4 Starch hydrolysis

Starch agar is used to test the ability of an organism to produce specific starch-hydrolyzing enzymes, that is, amylases. Starch agar is a simple nutritive medium containing fixed amount of starch [48]. Microbial cells hydrolyze the starch and make the zone free of starch. Iodine is added to the plate and incubated for specific period of time. Iodine turns blue, purple or black (depending on the concentration of iodine) in the presence of starch. A clear zone of hydrolysis around the microbial growth indicates the hydrolysis of starch by microbe.

6.3.5.5 MR-VP tests

Methyl red (MR) and Voges–Proskauer (VP) broth are used for performing MR and VP tests. The simple broth media contains peptone, buffers and dextrose or glucose. Different bacteria transform dextrose and glucose to pyruvate. Some pathways produce unstable acidic products which are immediately converted into neutral compounds. Some organisms follow the butylene glycol pathway resulting in the production of neutral end products like acetoin and 2,3-butanediol [49]. Also, some organisms may use the mixed acid pathway and produces acidic end products like lactic, acetic and formic acid which are stable in nature. The "MR" test involves addition of pH indicator (MR) to a tube of MR-VP broth inoculated with culture. If the microbe produces stable acidic end products, it will produce an acidic environment in the medium [50]. In acidic conditions, the MR will remain red in color. MR imparts yellow color at pH 6.2 and above while red color at pH 4.4 and below. On the other hand, phenol red turns yellow below a pH of 6.8. Though, in some cases, these two pH indicators may give confusing results [51]. The VP test identifies microbes that explore the butylene glycol pathway and therefore, produces acetoin [52]. If the VP reagents with MR-VP broth are fermented by an organism (by using the butylene glycol pathway), the oxidation of acetoin takes place in the presence of potassium hydroxide (KOH) and produces diacetyl [53]. Due to the presence of creatine in the reagent, diacetyl reaction produces red color. Development of red color is considered as a positive result while copper color is a sign of negative result. The MR and VP tests are specifically helpful in the identification of the *Enterobacteriaceae*.

6.3.5.6 Triple sugar iron agar

Triple sugar iron (TSI) agar is based on lactose, sucrose, some amount of glucose (dextrose) and also ferrous sulfate. It is a differential media containing pH indicator, that is, phenol red. It is generally used to differentiate the enteric bacteria depending on their ability to reduce sulfur and to ferment carbohydrates [54]. Due to the presence of phenol red in fermentation broth, in case of fermentation of any of the three sugars present in the medium, the color of medium will convert into yellow [55]. If an organism ferments the dextrose only, it will be utilized within the first few hours of incubation. Thereafter, acid-producing reaction generally reverts in the aerobic areas of the slant, and therefore, the medium in these areas turns red indicating the alkaline conditions [56]. On the other side, anaerobic areas will not usually revert to an alkaline condition, and therefore, retain yellow color.

If an organism reduces sulfur and produces hydrogen sulfide gas, it will react with the iron and appears as a black precipitate. These precipitate can cover the acid or alkaline results. Sulfur reduction usually requires an acidic condition; therefore, presence of black colored precipitate reveals occurrence of fermentation [57]. If the fermentation results in production of gas, it may develop fissures in the medium. Also, the produced gas may raise entire slant above the bottom of the test tube.

6.3.5.7 Catalase test

Catalase breaks hydrogen peroxide (H_2O_2) and produces H_2O and O_2. H_2O_2 is commonly used as a disinfectant on wounds. The bubbling that appears is due to the generation of O_2 gas [58]. Addition of few drops of H_2O_2 onto the smear of microbial suspension may be used to detect the presence of catalase. Rapid evolution of O_2 in form of bubbles is sign of a positive result. A negative result does not produce bubbles or results in few scattered bubbles.

6.3.5.8 Peroxidase test

This test is performed to check the aerobic or anaerobic nature of bacterium. It detects the electron transport chain, the final stage of aerobic respiration and generally, oxygen is the main final electron acceptor in this system [59]. A small volume of culture is taken on a clean slide or test tube. One drop of reagent is added onto the culture. Positive reactions turn the bacterial smear violet to purple either immediately or within 10 to 30 s.

6.3.5.9 Citrate test

Simmon's citrate agar detects the capability of an organism to use citrate as major carbon source. The citrate agar media is based on sodium citrate (in form of sole carbon source), a sole nitrogen source in form of ammonium dihydrogen phosphate, and also other nutrients along with a pH indicator, bromothymol blue [60]. The test is element of the IMViC tests. Organisms capable of utilizing citrate (as sole carbon source) use the citrase enzyme or the citrate permease for transportation of the citrate into the cell. The organism also converts ammonium dihydrogen phosphate in ammonia and ammonium hydroxide. Production of these molecules results in alkaline conditions in the medium. At pH 7.5 or above, bromothymol blue develops royal blue. At a neutral pH, the bromothymol blue turns green, as evidenced by the un-inoculated media [61]. If the medium turns blue, the organism is stated as citrate positive. In case of citrate negative microbe, there is no color change. In some cases, citrate negative organisms may also show weak growth on surface of media, but generally, they will not produce a color change.

6.3.5.10 Urease test

Urease test use the urease broth. This media detects the ability of an organism to produce urease enzyme. This enzyme hydrolyses urea to ammonia and carbon dioxide. The broth generally contains two pH buffers along with urea, a very small amount of nutrients for the bacteria and pH-based indicator (phenol red). Phenol red gives yellow color in an acidic environment while pink color in an alkaline environment [62]. The urease-producing microbe degrades the urea in the broth and results in production of ammonia. The produced ammonia makes environment alkaline in nature and therefore, media turns pink in color.

Though many enteric bacteria can have the ability to hydrolyze the urea, only a few of them can show rapid degradation of the urea. Such bacteria are generally recognized as "rapid urease-positive" microorganisms. Members belonging to the genus *Proteus* are considered among these organisms [63].

6.3.5.11 Decarboxylation test

Decarboxylase broth detects the presence of enzyme decarboxylase. This enzyme performs the removal of carboxyl group from an amino acid. This broth is based on dextrose, nutrients, pyridoxal and the pH-based indicators that is bromocresol purple and cresol red. Bromocresol purple gives purple color at an alkaline pH while turns yellow at an acidic pH [64]. In decarboxylase media, three amino acids are tested which includes arginine, lysine and ornithine. Three decarboxylase enzymes namely arginine decarboxylase, ornithine decarboxylase and lysine decarboxylase are routinely tested. Decarboxylation of the amino acid results in alkaline pH. Arginine is hydrolyzed to ornithine and further decarboxylated. Ornithine decarboxylation yields putrescine. Lysine decarboxylation results in cadaverine. These by-products significantly increase the pH of the media and therefore, the broth turns orange/yellow to purple [65]. If the media retains yellow color, or if there is no change in color is observed, the organism is generally considered decarboxylase negative for that amino acid. If the medium turns purple, the organism is considered decarboxylase positive for that amino acid [66].

6.4 Mycology

Mycology involves the study of fungi that includes both molds and yeasts. Mycology covers a large number of microbes [67]. Fungi are eukaryotic microorganisms which can produce the daughter fungi by division [68] or via sexual reproduction. Identification of new fungi, classification and their grouping with the existing fungi are some of the important aspects of mycology [69]. Fungi can be utilized for commercial purposes also. For example, some fungi have been reported to produce antibiotics, enzymes of commercial use, polysaccharides, pigments and other industrial relevant products [70]. Mycology covers all aspects related to fungi, their applications and economic importance [71].

6.4.1 Staining methods

6.4.1.1 The lactophenol cotton blue

This method of staining involves preparation of wet mount of fungi and subsequent staining observation of fungi.

Lactophenol cotton blue (LPCB) is a stain widely used for developing semipermanent microscopic preparation of fungi [72]. The LPCB stain has following three components:

Phenol: For killing and inhibition of any live organism
Lactic acid: For preservation of fungal structures
Cotton blue: To stain the cellulose and chitin content of the fungal cell wall intensely blue.

Procedure
1. Place one drop of 70% ethanol (v/v) on a glass slide.
2. Immerse the sample in the drop of alcohol.
3. Add 1–2 drops of the LPCB.
4. Cover the sample with coverslip gently avoiding formation of air bubbles.
5. Observe the slide by using suitable microscope.

6.4.2 KOH staining

Potassium hydroxide (KOH) based staining is generally used for the fast detection of fungus present in various clinical specimens [73]. The stain can be used to stain various specimens including hairs, skin, nails or sputum. The specimen is treated with KOH (20% w/v) to digest, soften and clear the tissues surrounding the fungi. Therefore, hyphae and spores of fungi can be seen under microscope.

Procedure
1. Take drop of KOH stain solution on a clean slide.
2. To this drop of KOH, transfer the specimen.
3. Cover the sample (with glass cover slip).
4. Place the slide in a clean petri dish, use a piece of filter paper or cotton wool to prevent the drying of preparation.
5. Examine the sample with help of microscope as soon as the specimen got cleared

6.4.3 Calcofluor white stain

"Calcofluor white" stain can be used for observing the yeasts, fungi and also some parasitic organisms. It is a non-specific fluorochrome which interacts and binds with cellulose and chitin present in the cell walls of various fungi and some other cellulose-containing organisms [74]. This staining helps in detecting the various yeasts and pathogenic fungi. The stain can also be mixed with potassium hydroxide mixture for clearing the specimen/sample [75]. Evans blue is a counterstain used to lessen the background fluorescence of tissues and cells while using the blue light excitation (not UV). The stained fungal or parasitic elements appear bright apple-green in color while other biological materials fluoresce reddish-orange while.

Procedure
Take the specimen onto a glass slide. Add a drop of "calcofluor white stain" and one drop of KOH (10%). Put a coverslip on specimen and allow the specimen to stain for 1 min. Now, cover the slide with a paper made towel and gently press to remove traces of excess fluid. Observe the slide.

Results
The fungi, pneumocystis cysts and parasites give a brilliant apple-green fluorescence with ultraviolet, violet and blue light [76]. In fluorescence microscopy, the green color is usually due to the use of barrier filters. The tissue samples may give yellowish-green background fluorescence. Though, the fluorescence shown by fungal or parasitic part is comparatively more intense and highly visible [77].

6.4.4 India ink staining

It is a negative staining technique. The organisms are observed as clear disc against a dark/black background. It is used for detection of capsule-producing fungus [78]. India ink can be used for samples like spinal fluids or exudates to provide a dark background that highlight hyaline yeast cells and capsular material (halo effect).

6.5 Parasitology

Parasitology in general includes the study of three major groups of organism, that is, parasitic protozoa, worms (parasitic helminths) and also those arthropods that directly cause disease or serve as vector of various pathogens [79]. Human infections caused by parasites range from relatively mild to fatal [79]. It has been observed that the incidence of some parasitic diseases have increased in recent years.

6.5.1 Assay of parasites

The malaria parasite can be detected and confirmed by microscopic tests.

6.5.1.1 Microscopic tests

The microscopic tests generally involve the preparation of sample, staining and direct visualization of the parasite with the help of microscope. Direct visualization of the malaria parasite through blood smears has been accepted as routine method for the diagnosis of malaria in most cases [80]. The most commonly used microscope based tests are the peripheral smear study and the quantitative buffy coat test. The light microscopy involving stained blood smears is the common and standard method for detection of the malaria parasite. The process involves preparation of blood smear and staining with specific Romanowsky stains. The red blood cells are examined in stained specimen for detection of malarial parasites [81]. The peripheral blood smear provides information on the species, the stages, and also the density of parasitemia [82]. Thick smear is used to detect infection and to estimate parasite concentration. The thick smear is usually dried for 30 min. It is not fixed with methanol. Due to the hemolysis of red blood cells and slow drying of specimen, the morphology of plasmodia may get adversely affected. On the other hand, thin smear is air dried for 10 min and fixed in methanol.

6.5.1.1.1 Staining
A number of Romanowsky stains (Field's, Wright's, Giemsa's and Leishman's) are considered for staining the prepared smears. Thick films are generally stained with the rapid Field's technique or Giemsa's stain for screening of parasites [83]. Thin blood films stained by Giemsa's or Leishman's stain are useful for specification of parasites**.**

6.5.1.1.2 Modified acid-fast staining procedure
It is used in identifying the oocysts of coccidian species (*Cryptosporidium, Cystoisospora* and *Cyclospora*), that may be tough to detect with other routine stains [84].

Sample for staining
Fresh (concentrated sediment) or preserved (formalin) stool may be used as sample. Other clinical specimens may also be used such as duodenal fluid, bile and pulmonary samples (induced sputum, bronchial wash and biopsies).

Reagents required: Absolute methanol, acid alcohol (prepared by mixing 10 mL of sulfuric acid and 90 mL of absolute ethanol): stored at room temperature, Kinyoun's carbol fuchsin, 3% malachite green.

Procedure

1. Prepare a smear with 1–2 drops of specimen on a slide. Dry using a slide warmer at 60 °C.
2. Fix specimen with help of absolute methanol (for 30 s).
3. Staining with Kinyoun's carbol fuschin (for 1 min) followed by rinsing with distilled water. Drain off extra liquid.
4. Destaining with acid alcohol (for 2 min). Rinse again using distilled water and drain extra liquid.
5. Counterstain the sample with malachite green (for 2 min). Repeat the rinse step with distilled water and drain off extra liquid.
6. Dry specimen using a slide warmer for around 5 min at 60 °C. Mount with a cover slip using desired mounting media.
7. Examine to confirm internal morphology.

Result

Cryptosporidium species stains a pinkish-red color. The background should observe as uniformly green stained.

6.5.1.1.3 Leishman's stain

Preparation of Leishman's stain solution

To prepare the stain, weigh and dissolve the 0.15 g of eosin methylene blue in methanol (100 mL). Dissolve the satin completely and remove the solution from the heater. Filter to clear the solution by using a dry Whatman paper filter [85] after cooling of solution up to the room temperature. Collect the clear filtered solution and store in a clean and dry brown glass bottle. Keep the solution at least 2–3 days before using it the first time. Store the Leishman's stain solution (at room temperature) in a tightly sealed bottle and protect it from light and heat.

Procedure

1. Prepare the thin smear and air-dry the smear.
2. Fully cover the smears with Leishman's stain solution for 2 min.
3. Add twice the amount of distilled water and gently mix by swirling. Incubate for 10 min.
4. Wash the slide with distilled water followed by drying of slide using blotting paper and air-dry.
5. Observe the slide under microscope.

Result

Cytoplasm of parasite: light blue in color while nuclei of parasite appears deep blue to blue-violet in color.

References

[1]   Kawahara K, Effects of heat fixation on tissue structure, immunostaining and in situ RT-PCR.J, Osaka Dent Univ, 1997, 31, 1–2.

[2]   Holger V, Heidrun P, Wolfgang, Immuno histochemical localization of P-glycoprotein in rat brain and detection of its increased expression by seizures are sensitive to fixation and staining variables, J Histochem Cytochem, 2005, 53, 517.

[3]   Penney DP, Frank M, Fagan C, Willis C, Stain and dye stability over a 30-year period: A comparison of certified dye powders by the biological stain commission, Biotech Histochem,2009, 84 (1), 11–15.

[4]   Robinow C, Kellenberger E, The bacterial nucleoid, Microbiol Mol Biol Rev, 1994, 58(2), 211–32.

[5]   Hobot JA, Villiger A, Escaig A, Maeder M, Ryter A, Kellenberger E, Shape and fine structure of nucleoids observed on sections of ultrarapidly frozen and cryosubstituted bacteria, J Bacteriol, 1985, 162(3), 960–71.

[6]   Lindsay MR, Webb RI, Strous M, Jetten MS, Butler MK, Forde RJ, Fuerst JA, Cell compartmentalisation in planctomycetes: Novel types of structural organisation for the bacterial cell, ArchMicrobiol, 2001, 175(6), 413–29.

[7]   Cabeen MT, Wagner CJ, Bacterial cell shape, Microbiol, 2005, 3, 601–10.

[8]   Kopito RR, Aggresomes, inclusion bodies and protein aggregation, Trends Cell Biol, 2000, 10(12), 524–30.

[9]   Sekula SB, Smits HL, Gussenhoven GC, Leeuwen JV, Amador S, Fujiwara T, Klatser PR, Oskam L, Simple and fast lateral flow test for classification of leprosy patients and identification of contacts with high risk of developing leprosy, J Clin Microbiol, 2003, 41(5), 1991–95.

[10] Moyes RB, Reynolds J, Breakwell DP, Differential staining of bacteria: Gram stain, CurrProtocMicrobiol, 2009, 10.1002/9780471729259.mca03cs15.

[11] James WB, Tod M, The gram stain, Bacteriol Rev, 1952, 16(1), 1–29.

[12] Grohmann E, Muth G, Espinsa M, Conjugative plasmid transfer in gram-positive bacteria, Microbiol Mol Biol Rev, 2003, 67(2), 277–301.

[13] Davies JA, et al., Chemical mechanism of the gram stain and synthesis of a new electron-opaque marker for electron microscopy which replaces the iodine mordant of the stain, J Bacteriol, 1983, 156(2), 837–45.

[14] Vairo MLR, Borzani W, Quantitative study of crystal violet adsorption by dead gram-negative bacteria, J Bacteriol, 1960, 80(4), 572.

[15] Beveridge TJ, Graham LL, Surface layers of bacteria, Microbiol Rev, 1991, 55(4), 684–705.

[16] Hussey MA, Zayaitz A, Acid-Fast Stain Protocols. American Society for Microbiology, Washington, DC. USA, 2008.

[17] Bishop PJ, Neumann G, The history of the Ziehl-Neelsen stain, Tubercle, 1970, 51, 196–206.

[18] Brennan PJ, Structure, function, and biogenesis of the cell wall of *Mycobacterium tuberculosis*, Tuberculosis, 2003, 83, 91–97.

[19] Breakwell DP, Moyes RB, Reynolds J, Differential staining of bacteria: Capsule stain, CurrProtoc Microbiol, 2009a, 10.1002/9780471729259.mca03is15.

[20] Griffith F, The significance of pneumococcal types, J Hyg, 1928, 27, 113–59.

[21] Gilbert I, Dissociation in an encapsulated staphylococcus, J Bacteriol, 1931, 21, 157–60.

[22] Hughes RB, Smith AC, Capsule Stain Protocols, USA, American Society for Microbiology, 2007.

[23] Gerhardt P, Murray RGE, Costilow RN, Nester EW, Wood WA, Krieg NR, Phillips GB, Manual of Methods for General Microbiology, Washington, DC, ASM Press, 1981.

[24] Gerhardt P, Murray RGE, Wood WA, Krieg NR, Methods for General and Molecular Bacteriology, Washington, DC, ASM Press, 1994.

[25] Moat GA, Foster JW, Microbial Physiology, 3rd Ed, 1995, New York, NY, Wiley-Liss Inc.

[26] Andrew B, Onderdonk DL, Kasper RL, Bartlett CG, The capsular polysaccharide of *Bacteroides fragilis* as a virulence factor: Comparison of the pathogenic potential of encapsulated and unencapsulated strains, J Infect Dis, 1977, 136, 82–89.

[27] Butt EM, Bonynge CW, Joyce RL, The Demonstration of Capsules about Hemolytic*Streptococci* with India Ink or Azo Blue, J Infect Dis, 1936, 58, 5–9.

[28] Schaeffer AB, Fulton M, 1933, A simplified method of staining endospores Science, 77, Issue (1990), 194, 10.1126/science.77.1990.194.

[29] Harley and Prescott: Laboratory Exercises in Microbiology, 2002, McGraw Hill, Vol. 58.

[30] Gerhardt P, Ribi E, Ultrastructure of the exosporium enveloping spores of *Bacillus cereus*, J Bacteriol, 1964, 88(6), 1774–89.

[31] Rouf MA, Stokes JL, Isolation and identification of the sudanophilic granules of *Sphaerotilusnatans.*, J Bacteriol, 1962, 83(2), 343–47.

[32] Widra A, Metachromatic granules of microorganisms, J Bacteriol, 1959, 78(5), 664–70.

[33] Kumar S, Essentials of Microbiology. ISBN 978-93-5152-380-2, 2016, India, Jaypee Brothers Medical Publishers.

[34] Breakwell DP, Moyes RB, Reynolds J, Differential staining of bacteria: Flagella stain, CurrProtocMicrobiol, 2009b, 10.1002/9780471729259.mca03gs15.

[35] Geis G, Leying H, Suerbaum S, Mai U, Opferkuch W, Ultrastructure and chemical analysis of *Campylobacter pylori* flagella, J Bacteriol, 1989, 27(3), 436–41.

[36] Lowy J, Hanson J, Electronmicroscope studies of bacterial flagella, J Mol Biol, 1965, 11(2), 293–313.

[37] Leifson E, Staining, shape, and arrangement of bacterial flagella, J Bacteriol, 1951, 62, 377–89.

[38] Pommerville JC, Alcamo's Laboratory Fundamentals of Microbiology, 2004, Sudbury, Jones & Bartlett Publishers.

[39] Tittsler RP, Sandholzer LA, The use of semi-solid agar for the detection of bacterial motility, J Bacteriol, 1936, 31(6), 575–80.

[40] Pfennig N, Wagener S, An improved method of preparing wet mounts for photomicrographs of microorganisms, J Microbiol Methods, 1986, 4(5–6), 303–06.

[41] Peskin CS, Odell GM, Oster GF, Cellular motions and thermal fluctuations: The Brownian ratchet, Biophys J, 1993, 65(1), 316–24.

[42] Macnab RM, Examination of bacterial flagellationby dark-field microscopy, Clin Microbiol, 1976, 4(3), 258–65.

[43] Leifson E, Determination of carbohydrate metabolism of marine bacteria, J Bacteriol, 1963, 85(5), 1183–84.

[44] Graham DC, Hodgkiss W, Identity of gram negative, yellow pigmented, fermentative bacteria isolated from plants and animals, J ApplMicrobiol, 1967, 30(1), 175–89.

[45] Griffiths MW, Phillips JD, Muir DD, Thermostability of proteases and lipases from a number of species of psychrotrophic bacteria of dairy origin, J ApplMicrobiol, 1981, 50(2), 289–303.

[46] Gupta R, Gupta N, Rathi P, Bacterial lipases: An overview of production, purification and biochemical properties, ApplMicrobiolBiotechnol, 2004, 64(6), 763–81.

[47] Sztajer H, Maliszewska I, Wieczorek J, Production of exogenous lipases by bacteria, fungi, and actinomycetes, Enzyme MicrobTechnol, 1988, 10(8), 492–97.

[48] Macfarlane GT, Englyst HN, Starch utilization by the human large intestinal microflora, J ApplMicrobiol, 1986, 60(3), 195–201.

[49] Barritt Maxwell M, The intensification of the Voges-Proskauer reaction by the addition of α-naphthol, J Pathol, 1936, 42(2), 441–54.

[50] Levine M, The correlation of the Voges-Proskauer and methyl-red reactions in the colonaerogenes group of bacteria, J Infect Dis, 1916, 18(4), 358–67.

[51] Twedt RM, Spaulding PL, Hall HE, Morphological, cultural, biochemical, and serological comparison of japanese strains of *Vibrio parahemolyticus* with related cultures Isolated in the United States, J Bacteriol, 1969, 98(2), 511–18.

[52] Paine FS, The destruction of acetyl-methyl-carbinol by members of the colon-aerogenes group, J Bacteriol, 1927, 13, 269–74.

[53] Brenner DJ, McWhorter AC, Knutson JKL, Steigerwalt AG, *Escherichia vulneris*: A new species of *Enterobacteriaceae* associated with human wounds, J Clin Microbiol, 1982, 15, 1133–40.

[54] Durand AM, Blazevic DJ, Differentiation of *Serratia* from *Enterobacter* on the basis of nucleoside phosphotransferase production, Appl Environ Microbiol, 1970, 19(1), 134–37.

[55] Janda JM, Abbott SM, Cheung WK, Hanson DF, Biochemical identification of citrobacteria in the clinical laboratory, J Clin Microbiol, 1994, 32(8), 1850–54.

[56] Archer JR, Romero S, Ritchie AE, Hamacher ME, Steiner BM, Bryner JH, Schell RF, Characterization of an unclassified microaerophilic bacterium associated with gastroenteritis, J Clin Microbiol, 1988, 26, 101–05.

[57] Brooker DC, Lund ME, Blazevic DJ, Rapid test for lysine decarboxylase activity in *Enterobacteriaceae*, ApplMicrobiol, 1973, 26, 622–23.

[58] Taylor WI, Achanzar D, Catalase Test as an Aid to the Identification of *Enterobacteriaceae*, Appl Environ Microbiol, 1972, 24(1), 58–61.

[59] Pine L, Hoffman PS, Malcolm GB, Benson RF, Gorman GW, Whole-cell peroxidase test for identification of *Legionella pneumophila*, J Clin Microbiol, 1984, 19(2), 286–90.

[60] Graevenitz AV, Practical substitution for the indole, methyl red, Voges-Proskauer, citrate system, Appl Environ Microbiol, 1971, 21(6), 1107–09.

[61] Simmons JS, A culture medium for differentiating organisms of typhoid-colon aerogenes groups, J Infect Dis, 1926, 39(3), 209–14.

[62] Vaira D, Holton J, Cairns S, Polydorou A, Falzon M, Dowsett J, Salmon PR, Urease tests for *Campylobacter pylori*: Care in interpretation, J Clin Pathol, 1988, 41(7), 812–13.

[63] Westblom TU, Madan E, Kemp J, Subik MA, Evaluation of a rapid urease test to detect *Campylobacter pylori* infection, J Clin Microbiol, 1988, 26(7), 1393–94.

[64] Carlquist PR, A biochemical test for separating paracolon groups, J Bacteriol, 1956, 71, 339–41.

[65] Pegg AE, Regulation of ornithine decarboxylase, J Biol Chem, 2006, 281, 14529–32.

[66] Fecker LF, Rügenhagen C, Berlin J, Increased production of cadaverine and anabasine in hairy root cultures of *Nicotiana tabacum* expressing a bacterial lysine decarboxylase gene, Plant Mol Biol, 1993, 23, 11–12.

[67] Riddell RW, Permanent stained mycological preparations obtained by slide culture, Mycologia, 1950, 42(2), 265–70.

[68] Heitman J, Sun S, James TY, Evolution of fungal sexual reproduction, Mycologia, 2013, 105(1), 1–27.

[69] Dw R, Rj L, *Metarhizium spp.*, Cosmopolitan insect-pathogenic fungi: Mycological aspects, Adv ApplMicrobiol, 2004, 54, 1–70.

[70] Allan CR, Hadwiger LA, The fungicidal effect of chitosan on fungi of varying cell wall composition, Exp Mycol, 1979, 3, 285–87.

[71] Cueto M, Jensen PR, Kauffman C, Fenical W, Lobkovsky E, Clardy JP, A new antibiotic produced by a marine fungus in response to bacterial challenge, J Nat Prod, 2001, 64, 1444–46.

[72] Parija SC, Prabhakar PK, Evaluation of lacto-phenol cotton blue for wet mount preparation of feces, J Clin Microbiol, 1995, 33(4), 1019–21.

[73] Vierheilig H, Coughlan AP, Wyss U, Piché Y, ink and vinegar, a simple staining technique for arbuscular-mycorrhizal fungi, Appl Environ Microbiol, 1998, 64(12), 5004–07.

[74] Rasconi S, Jobard M, Jouve L, Ngando TS, Use of calcofluor white for detection, identification and quantification of phytoplanktonic fungal parasites, Appl Environ Microbiol, 2009, 75(8), 2545–53.

[75] Roncero C, Durán A, Effect of calcofluor white and congo red on fungal cell wall morphogenesis: In vivo activation of chitin polymerization, J Bacteriol, 1985, 163(3), 1180–85.

[76] Monheit JG, Brown G, Kott MM, Schmidt WA, Moore DG, Calcofluor white detection of fungi in cytopathology, Am J Clin Pathol, 1986, 85(2), 222–25.

[77] Stratton N, Hryniewicki J, Aarnaes SL, Tan G, Maza LM, Peterson EM, Comparison of monoclonal antibody and calcofluor white stains for the detection of *Pneumocystis carinii* from respiratory specimens, J Clin Microbiol, 1999, 29(3), 645–47.

[78] Avital ERJ, Abadi M, Immune reconstitution cryptococcosis after initiation of successful highly active antiretroviral therapy, Clin Infect Dis,2002, 35, 128–33.

[79] Jennings JB, Parasitism and Commensalism in the *Turbellaria*, Adv Parasitol, 1971, 9, 1–32. commensalism, colonization, infection, and disease.InfectImmun 2000, 68, 6511–6518.

[80] Foggie A, Studies on the infectious agent of tick-borne fever in sheep, J Pathol, 1951, 63, 1–15.

[81] Wittekind D, On the nature of Romanowsky dyes and the Romanowsky-Giemsa effect, Int J Lab Hematol, 1979, 1(4), 247–62.

[82] Lima MR, Eisen H, Minoprio P, Joskowicz M, Coutinho A, Persistence of polyclonal B cell activation with undetectable parasitemia in late stages of experimental Chagas' disease, J Immunol, 1986, 137(1), 353–56.

[83] Warhurst DC, Williams JE, Laboratory diagnosis of malaria, J Clin Pathol, 1996, 49(7), 533–38.

[84] Lambros C, Vanderberg JP, Synchronization of *Plasmodium falciparum* erythrocytic stages in culture, J Parasitol, 1979, 65(3), 418–20.

[85] Hughes DE, A press for disrupting bacteria and other micro-organisms, Br J Exp Pathol, 1951, 32(2), 97–109.

Anil Kumar, Shailja Sankhyan, Abhishek Walia,
Chayanika Putatunda, Dharambir Kashyap, Ajay Sharma
and Anil K Sharma

Chapter 7
Antimicrobial resistance: medical science facing a daunting challenge

Abstract: Antimicrobial resistance (AMR) is the potential of the microorganisms to resist the impacts of medicine/drugs that once could positively treat the microbial infection. The resistant microorganisms are hard to treat, may require elevated antimicrobial doses or alternative medications. Resistance arises through several mechanisms including natural resistance, by one species acquiring resistance from another or genetic mutation. All types of microorganisms including fungus, viruses, bacteria and protozoa can develop resistance. The increasing resistance among microbes is a major concern among medical microbiologists and other researchers related to the field. This chapter covers the AMR development among microbes including possible mechanisms, biochemical and molecular cascades.

Keywords: microbes, antimicrobial resistance, genetic mutation, mechanism of antimicrobial resistance

7.1 Introduction

The potential of microbes to attain resistance has been well known soon after the discovery of penicillin by Alexander Fleming, and today resistance has developed to almost every antibiotic launched. In the golden era (1940–1990) of antibiotics, numerous

Anil Kumar, Department of Microbiology, DAV University, Jalandhar, Punjab, India,
e-mail: anilsharma2710@gmail.com
Shailja Sankhyan, Department of Biotechnology, Chandigarh University Gharuan, Mohali, India
Abhishek Walia, Department of Microbiology, CSK Himachal Pradesh Agricultural University,
Palampur, India
Chayanika Putatunda, Department of Microbiology, DAV University, Jalandhar, Punjab, India
Dharambir Kashyap, Department of Histopathology, Postgraduate Institute of Medical Education
and Research, Chandigarh 160 012, Punjab, India
Ajay Sharma, Department of Chemistry, Career Point University, Tikker-Kharwarian, Hamirpur
176 041, Himachal Pradesh, India
Anil K Sharma, Department of Biotechnology, Maharishi Markandeshwar (Deemed to be
University) Mullana, Ambala, Haryana, India, e-mail: anibiotech18@gmail.com

https://doi.org/10.1515/9783110517736-007

classes including aminoglycosides, tetracyclines, glycopeptides, polymyxins, mac-rolides, β-lactam antibiotics and fluoroquinolones were developed [1, 2]. However, the effective role of antibiotics addressing bacterial infections was taken for granted because of which new microbes have emerged that are capable of destroying the antibiotics negating its effects. This has significantly increased the mortality, mor-bidity and healthcare cost [2, 3].

Antimicrobial resistance (AMR) is the potential of the microorganisms to resist the impacts of medicine/drugs that once could positively treat the microbial infection [4]. The resistant microorganisms are hard to treat, requiring elevated antimicrobial doses or alternative medications but may be more expensive and toxic. Resistance arises through several mechanisms including natural resistance, by one species ac-quiring resistance from another, or genetic mutation [5]. All types of microorganisms including fungus, viruses, bacteria and protozoa can develop resistance. The increase in the AMR is stemmed due to number of factors: (i) improper use of antibiotics in patients, (ii) extensive use in animal industry, (iii) suboptimal prescription and (iv) incompletion of the prescribed antibiotic course [6, 7]. The continuous rise in the an-tibiotic resistance is of serious concern and the "superbugs" emerged now have been found to be resistant to numerous antibiotics. These include the ESKAPE pathogens (*Enterococcus faecium*, *Staphylococcus aureus*, *Klebsiella pneumoniae*, *Acinetobacter baumannii*, *Pseudomonas aeruginosa* and *Enterobacter* sp.) (Table 7.1). Globally, 700,000 patients die per annum because of AMR and have been estimated to increase up to 10 million by 2050 [8]. This poses a challenge in the field of medicine, world-wide. The increased and widespread AMR, emergence of new pathogens and endless evolution are driving the quest for novel drugs, a necessity of the hour.

7.2 Molecular cascade in antimicrobial resistance

AMR among bacteria is a complex phenomenon associated with various mecha-nisms. The bacterial species employ various strategies to avoid antibiotic effects. The mechanism of acquisition of antibiotic resistance among bacteria is a very com-plex process [16]. Their integrated network or cascade of antibiotic resistance ele-ments seem to provide protection to bacteria from various antimicrobial chemicals. The front line of defense among bacteria is cell wall envelope. The outer cell wall of bacteria is a very complex structure that provides effective protection against anti-microbial substances [17]. In gram-negative (–ve) bacteria, the LPS and porins are known to provide selective resistance against antimicrobial substances. Further, the genome of these bacteria contains genes which encode small transport proteins and these proteins make efflux and influx system in bacteria. So, the gene expres-sion selectively controls the expression of efflux pumps which in turn confirms re-sistance against antibiotics. Thus, by controlling antibiotics influx and efflux, they

Table 7.1: Microorganisms resistant to drugs [9–15].

Sr. No.	Microorganisms	Resistant drugs
Bacterium		
1.	*Escherichia coli*	Fluoroquinolones and cephalosporins
2.	*Streptococcus pneumoniae*	Penicillin
3.	*Klebsiella pneumoniae*	Cephalosporins and carbapenems
4.	*Mycobacterium tuberculosis*	Rifampicin, isoniazid and fluoroquinolone
5.	*Neisseria gonorrhoeae*	Cephalosporins and fluoroquinolone
6.	*Staphylococcus aureus*	Methicillin and vancomycin
7.	*Shigella species*	Fluoroquinolones
8.	*Salmonella* sp.	Fluoroquinolones
9.	*Acinetobacter baumannii*	Carbapenem
10.	*Pseudomonas aeruginosa*	Carbapenem
11.	*Enterococcus faecium*	Vancomycin
12.	*Helicobacter pylori*	Clarithromycin
13.	*Campylobacter* sp.	Fluoroquinolone
14.	*Haemophilus influenzae*	Ampicillin
Fungus		
15.	*Candida* sp.	Fluconazole and echinocandins
16.	*Aspergillus* sp.	Azoles
17.	*Cryptococcus neoformans*	Fluconazole
18.	*Scopulariopsis* sp.	Amphotericin B, flucytosine and azoles
Viruses		
19.	Hepatitis B virus (HBV)	Lamivudine
20.	Human immunodeficiency virus (HIV)	Antiretroviral drug
21.	Cytomegalovirus (CMV)	Ganciclovir and foscarnet
22.	Herpes simplex virus (HSV)	Acyclovir, valacyclovir and famciclovir
23.	Influenza virus	Neuraminidase inhibitors and adamantane derivatives (amantadine and rimantadine)

control the effective concentration of antibiotics and therefore affect the antibiotics binding with cellular targets, thus contribute to resistance [18, 19]. Further efflux and influx system can be co-regulated and have a connection with other genetic network that codes auxiliary proteins [20]. This co-regulation has been discovered in *Escherichia coli* mar locus regulon. The exposure of *E. coli* to antibiotics gives rise to diverse effects that range from efflux gene expression to alteration in porin protein contents and change in the expression of several stress genes [18, 21]. There are a different set of efflux pumps associated with antibiotic resistance (1) ATP binding cassette ABC, (2) MATE-multidrug and toxic efflux, (3) SMR-small multidrug resistance (MDR), (4) RND-resistance nodulation division and (5) MF-major facilitator [22]. Other resistance systems are more specific in bacteria that range from enzyme inactivation of antibiotics (β-lactamases) to target modification and resistance genetics element mobilization between bacteria. Here in the current chapter, we discuss the target modification (point mutations) and mobile genetic elements based molecular mechanisms of antibiotic resistance [20].

Gene mutation is the one phenomenon where point mutations alter the structural or regulatory genes that may reduce the binding affinity of antimicrobials; therefore, resulting in resistance phenotype [20]. For example, gene mutation which encodes the DNA gyrase can affect the binding efficacy of quinolone. Further, the multiple gene mutations which encode the targets for antimicrobial compounds may lead to a higher level of resistance among microbes. Such as in DNA topoisomerase genes, gyrA, gyrB and parC, the manifold mutations in these genes have elevated the values of minimum inhibitory concentrations (MICs) thus reducing the bacterial susceptibility to quinolone antibiotics [23]. Further, the multiple point mutations in β-lactamase genes resulted in identification of more than 300 β-lactamases which are associated with resistance phenotypes [24, 25]. Apart from the point or multiple mutations, the acquirement of mobile DNA components augmented the antibiotic resistance among bacteria. The transfer of these gene components through transposons, plasmids and integrons has extended the resistant microbes of human or veterinary importance [16]. Bacterial plasmids are one of the primary sources to acquire resistance among bacterial species. They are extra-chromosomal genetic elements having ability to replicate independent chromosomal DNA, and they usually transfer among bacteria using horizontal transfer mechanisms. These extrachromosomal DNA are not crucial for bacterial survival but carry the genes which are required for virulence, adhesion and AMR [25]. Those plasmids which carry the antibiotic resistance genes are called R plasmids. R plasmids significantly contributed to the resistance phenotypes for not only quinolones but also for β-lactam antibiotics. Previously, it has been believed that the quinolone antibiotic resistance is a chromosomal origin. However, with the discovery of plasmid-mediated quinolone resistance (qnr) in *Klebsiella*, it has been shown that plasmids can contribute to quinolone resistance. It has been observed that the qnr gene encodes a protein which blocks the interaction of ciprofloxacin with DNA topoisomerase IV and DNA gyrase [23, 26]. This resistance gene was discovered as an integron-like structure near Orf513 on the MDR plasmid, pMG252. Now, different qnr plasmids have been discovered from clinical isolates of *Salmonella* sp., *Klebsiella* sp., *Escherichia* sp., *Citrobacter* sp. and so on. The diverse qnr genes documented till date include qnrA, qnrB and qnrS [27–29].

Apart from holding resistance genes, plasmids play a significant role to carry forward another resistance element against antimicrobials such as transposons and integrons. Transposons are the genetic elements or gene sequences that can travel from chromosomes to chromosomes or plasmid. These elements consist of insertion sequences, intervening DNA and sequences that code an enzyme transposase [30]. These elements are also named as jumping genes, and they can be simple or complex like composite transposons. The composite transposons contain a key segment/region having genes other than that needed for transposition. The composite transposons contain genes resistant to antibiotic flanked on either sides by insertional elements [26]. Now, it has been evident that the large number of antibiotic resistant markers transfer among bacteria through transposons. These transposons usually transfer among different bacterial species using conjugation and are therefore called conjugative transposons. These transposons are a hybrid of plasmids and transposons. They are also known to form single-stranded closed DNA after excision from chromosome-like plasmids and then transfer to the neighboring bacterium through conjugation. After conjugation, they subsequently integrate into the host chromosomes or plasmids. Such type of transposons has been discovered in both gram-positive (−ve) or gram-positive (+ve) bacteria [31].

Another important molecular mechanism through which the bacteria attain antibiotic resistance genes is named integron. Integrons are also movable genetic components that contain a central region for antibiotic resistance (gene cassettes) flanked by conserved regions. These gene cassettes act as free circular DNA molecules that lack promoter regions. A 59bp sequence is located downstream that acts as recombination site in promoterless resistance genes. The insertion of antibiotic resistance cassette to integron through recombination at recombination site downstream of the promoter help the expression of inserted cassette. At a time, multiple gene cassettes can be organized in tandem that confirm the antibiotic resistance. Different classes of integrons have been identified and recognized on the bases of an integrase gene sequence, and among them, class I integrons are the most frequently found in bacteria. In addition to these mobile elements, the other molecular mechanisms by which the bacteria can attain resistance are the uptake of naked genetic material from adjacent environment (transformation) or during infection with bacteriophages (transduction) [26, 29, 30, 32]. Thus, these mechanisms have contributed significantly to the speedy dispersion of antibiotic resistance among bacteria not only in a natural environment but also in humans.

7.3 Biochemical aspects of antibiotic resistance

Antibiotic drug resistance has attracted immense importance in the past years as it hinders the process of treatment of infectious diseases. Development of resistance among the key microbial pathogens is a well-known public health threat; affecting human life globally. The response of bacteria toward the antimicrobial attack is the microbial adaptation and the starting of evolution. The "survival for the fittest" is a result of vast genomic flexibility of pathogenic bacteria that activate specific signals/responses, which results in mutational adaptations, genetic material attainment or gene expression alterations; creating resistance to the majority of antibiotics presently accessible in medical sector. Therefore, understanding the biochemical aspects of antibiotic resistance is of extreme importance to circumvent the occurrence/dispersal of resistance and to formulate/invent new therapeutic strategies antagonistic to MDR microbes [33, 34].

The way bacteria acquire resistance may vary among bacterial species, which comprises of the following mechanisms (Figure 1):

(i) Antibiotic inactivation – The antibiotic molecules are directly inactivated by enzymes produced by bacteria [34];

(ii) Target modification – The antibiotic sensitivity is changed by modifying or altering the drug target in the bacteria [35];

(iii) Antibiotic efflux and outer membrane (OM) permeability changes – There is a decrease in the uptake of antimicrobial compound thus preventing the drug from reaching the intracellular targets. The efflux system extrudes/pumps the antibiotic compound out from the cell and leads to the development of resistance [36];

(iv) Target bypass – Some bacteria bypass the inactivation of a given enzyme by becoming refractory to these antibiotics, as seen in many sulfonamide and trimethoprim-resistant bacteria. They bypass the inhibition of enzymes required for their inactivation, that is, dihydropteroate synthase and dihydrofolate reductase (involved in tetrahydrofolate biosynthesis), otherwise, inhibited by sulfonamides and trimethoprim, respectively. In this type of mechanism, the resistant bacteria secrets a second enzyme having low affinity toward the inhibitors [33, 37].

The antibiotic resistance mechanisms are diverse and a single bacterium may have different resistance mechanisms of the four mechanisms mentioned above. These mechanisms depend upon multiple factors such as the bacterial strain type, antibiotics nature, site targeted and the resistance interceded by mutations in plasmid/chromosome.

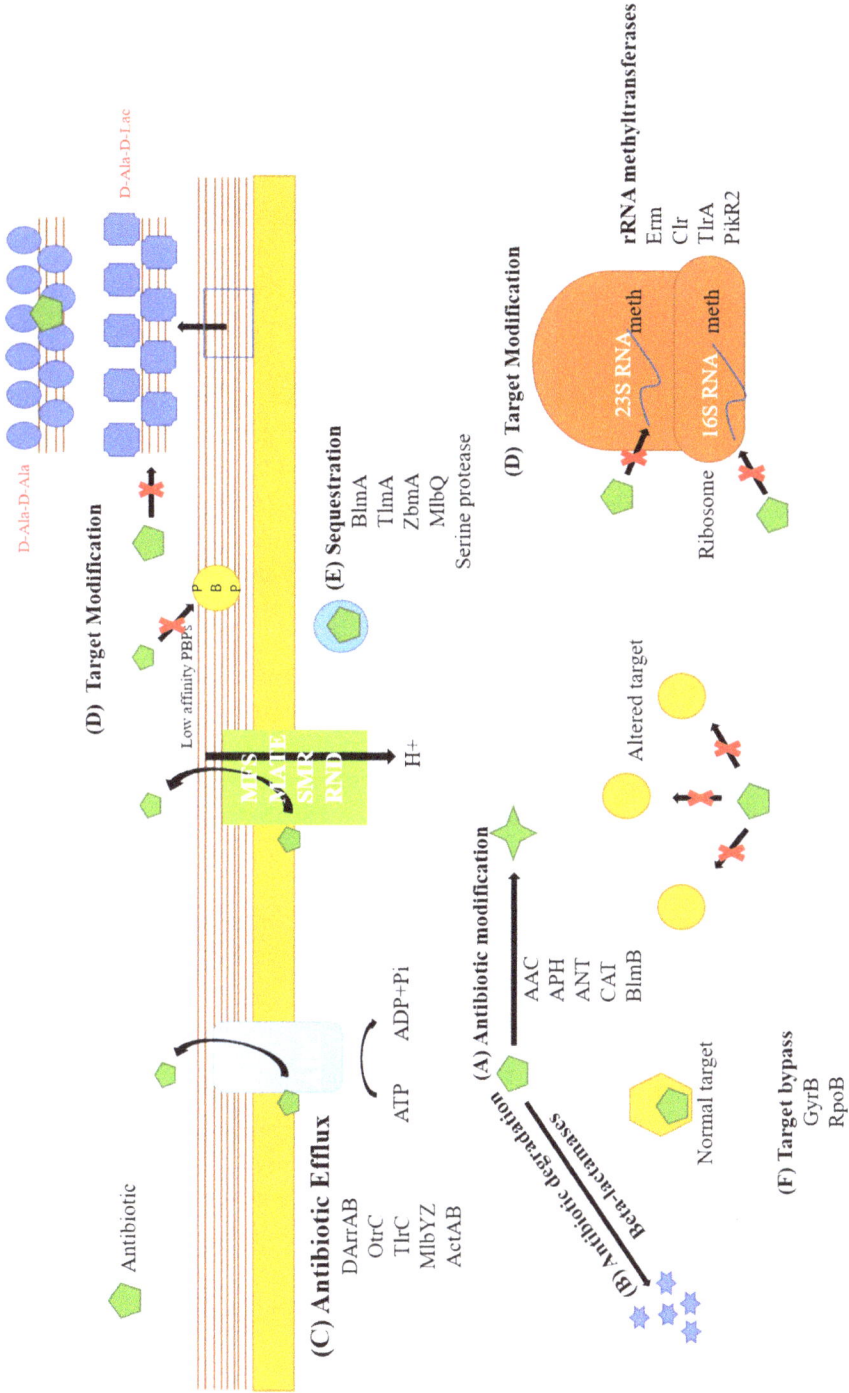

Figure 1: Representation of the bacterial antibiotic resistance mechanisms.

7.3.1 Antibiotic inactivation

The inactivation process involves secretion of those enzymes capable of destructing or altering the antibiotic itself. Biochemical approaches involve these three aspects, that is, hydrolysis, group transfer and redox mechanisms.

7.3.1.1 Antibiotic inactivation by hydrolysis

Several antibiotics possess chemical bonds that are hydrolytically susceptible, that is, esters and amides. Different groups of enzymes that target and cleave these chemical bonds to destroy/degrade antibiotic activity are known. These enzymes are produced/secreted by bacteria, prevents the antibiotic molecule to reach to their target within the bacteria and thereby inactivates the drug. The β-lactamases are the hydrolytic enzymes known to break the β-lactam ring of the antibiotics such as cephalosporin and penicillin. Numerous gram +ve and gram –ve bacteria are known to produce/secrete such enzymes and till date greater than (>) 200 distinct β-lactamases are known. The β-lactamases belonging to different functional groups based on functional characteristics and preference for particular antibiotic substrate have been explored [39, 40]. Extended-spectrum β-lactamases (ESBLs) showed resistance to antibiotic molecules such as penicillins, aztreonam and third-generation cephalosporins, but not to carbapenems and cephamycins. The ESBLs showed greater diversity in nature, and > 180 distinct ESBLs have been known. They are usually found in *K. pneumoniae*, *Proteus mirabilis* and *E. coli*, but also present in other members of Enterobacteriaceae [41, 42].

7.3.1.2 Antibiotic inactivation by group transfer

Transferases are the enzymes that inactivate antibiotics such as aminoglycosides, rifampicin and chloramphenicol by substituting/adding chemical group like acetyl, phosphory and adenylyl to the periphery of antibiotic compounds. These alterations affect the antibiotic binding to the target site. The chemical approaches used are thiol transfer, *O*-acetylation [43, 44], *O*-phosphorylation [45, 46], *O*-ribosylation [47], *O*-glycosylation and *N*-acetylation [43, 44]. These covalent modification approaches need a co-substrate for their activity (ATP, NAD⁺, acetyl-CoA, UDP-glucose) and are limited to the cytoplasm.

7.3.1.3 Antibiotic inactivation by redox process

Antibiotic inactivation also occurs by the oxidation and reduction process of the antibiotics. These types of processes are rarely used by pathogenic bacteria, for example, (i) the tetracycline oxidation by the enzyme TetX, (ii) *Streptomyces virginiae* that produces type A streptogramin antibiotic virginiamycin M1; reduces a critical ketone group to an alcohol at position 16 to protects itself from its own antibiotic.

7.3.2 Target modification

The alteration in antibiotic target sites is another most common mechanism of acquiring resistance, which makes the antimicrobial compound impotent to bind appropriately. Target site is essential for cellular functions, and microorganisms cannot avoid antimicrobial action completely. On the other hand, it is possible that mutations takes place in the target site which decrease the susceptibility to inhibition without altering the cellular functions [50]. In some cases, the resistance produced due to alteration in the target site require some additional cellular changes to compensate for the modified target characteristics. For example, *S. aureus* acquire and express foreign penicillin-binding protein (PBPs) and PBP2a and becomes resistance to methicillin and other β-lactam antibiotics but can coup transpeptidation (cross-linking) of host PBPs. To maintain the efficacy of peptidoglycan biosynthesis, PBP2a requires modifications in peptidoglycan composition and structure, which involves functioning of many additional genes [37, 51, 52].

7.3.2.1 Peptidoglycan structure alteration

Synthesis of peptidoglycan take place in three different stages. The first stage occurs in the cytoplasm, which involve the synthesis of low-molecular-weight precursors that is, UDP-GlcAc and UDP-MurNAc-L-ala-D-glu-meso-Dap-D-ala-D-ala. In the second stage, membrane-bound enzymes catalyze the synthesis of peptidoglycan. The third stage involves subunit polymerization and nascent peptidoglycan attachment to the cell wall in a transpeptidase reaction. The cell wall peptidoglycan is an excellent selective target for the action of antibiotic molecules. The peptidoglycan is essential for the growth/survival of majority of bacteria. Therefore, the enzymes engaged in peptidoglycan synthesis and assembly can be a prime target for selective inhibition of the bacterial cell wall. The PBPs on binding to a β-lactam antibiotic affects the morphological response of the bacterium toward the agent. For instance, few antibiotics bind to the PBPs that have role in septum formation during cell division, because of which bacteria continue to multiply in long filaments and ultimately die. In addition, some β-lactam antibiotic (mecillinam) do not bind to PBPs of gram +ve bacteria thereby letting the bacteria unaffected. The PBPs modification have induced penicillin and ampicillin resistance among *Streptococcus pneumoniae* and *E. faecium*, respectively [53, 54]. The PBP2a found in *S. aureus* is resistant to methicillin and all β-lactams antibiotics, but remains functional for cell wall synthesis even at fatal concentrations of β-lactam. The glycopeptide antibiotic-like vancomycin acquire resistance by modifying the target site from D-Ala-D-Ala (D-alanyl-D-alanine)) terminating muropeptide to D-Ala-D-Lac (D-alanyl-D-lactate) phenotype, which results in decrease of the vancomycin affinity for Lipid II [55, 56].

7.3.2.2 Protein synthesis interference

The vital role of bacterial ribosome makes it a preferred target for antimicrobial compounds. The broad range of clinically important antimicrobials interfere with complex translational protein machinery at different stages of protein metabolism. The resistance to peptide antimicrobials is developed by altering the specific target [37]. The antimicrobial compounds macrolide, lincosamide and streptogramin B (MLS(B)) binds to 50S ribosomal subunit and block the bacterial protein synthesis [57, 58]. The MSL (B) resistance occurs in a number of Gram −ve and Gram + ve bacteria, which results from a post-transcriptional modification of the 23S rRNA constituent of the 50S subunit of ribosomes [59]. This embroil methylation/dimethylation of essential adenine bases in the functional domain of peptidyl transferase. The resistance to macrolide group of antibiotics in a diversity of microorganisms has also been associated with mutational changes in 23S rRNA adjacent to the sites of methylation. However, the macrolide resistance in *S. pneumoniae* has been reported due to modifications in the L22 and L4 proteins of the 50S subunit in spite of multiple mutations in 23S rRNA subunit. The mutational changes in the 16S rRNA region showed resistance to aminoglycosides [60, 61]. The microorganisms known to produce aminoglycosides develop higher level of antibiotic resistance by 16S rRNA gene posttranscriptional methylation in the binding site of aminoglycosides. The initial uptake of aminoglycoside to the cytoplasmic membrane is energy dependent. Aminoglycosides which enter the cell will bind only to the ribosomes actively involved in synthesis of protein. The ribosomal binding induces a protein that has a role in aminoglycosides uptake. Aminoglycosides which are acetylated, phosphorylated or adenylated by the enzymes present in the periplasmic space do not bind efficiently to the ribosomes, and hence their uptake is poor and there is no induction of the transport protein [62].

7.3.2.3 DNA synthesis interference

Chloroquine are the intercalating agents which harm DNA template function. Nalidixic acid ciprofloxacin, ofloxacin and norfloxacin are inhibitors of DNA replication.

Rifampicin acts as an inhibitor of RNA polymerase. Fluoroquinolones also prevent DNA replication and transcription by interacting with the enzymes DNA gyrase and topoisomerase IV. The resistance occurs due to the mutational changes in precise segments of the structural genes that adequately modify these enzymes, hence prevents binding [63, 64]. The most frequent mutations occurring in this segment showed resistance via reduced drug affinity for the altered DNA–gyrase complex [65].

7.3.3 Efflux pumps and decreased membrane permeability

These efflux pumps are the transport proteins that expel/extrude antimicrobial drugs out from the cell and maintain its intracellular concentrations at very low levels. The decreased OM penetration leads to reduced uptake of antibiotics. The active efflux pump and reduced uptake bring resistance to low level in numerous important pathogenic bacteria [66].

7.3.3.1 Efflux pumps

Different classes of antimicrobials including tetracyclines and fluoroquinolones exert intracellular effects to inhibit microbial protein and DNA biosynthesis and thus are affected by efflux machinery. The efflux machinery differ in their action mechanism and specificity [67]. It may be drug specific or with broad drug specificity (proficient of extruding a range of structurally unrelated drugs), usually seen in MDR bacteria [68, 69]. The bacterial strains, both gram +ve and gram −ve, can have single-drug or multiple drug efflux system. For example, the efflux mechanism of tetracycline resistance lead to decreased drug accumulation. The decreased drug uptake and efflux happen at the same time. The bacteria resistant to tetracycline binds poorly, and the accumulated drug is pumped out by Tet efflux machinery using an energy-dependent process [70, 71]. The intrinsic AMR in a variety of organisms is because of inducible multidrug efflux pumps, and mutations in regulatory components controlling the production of efflux machinery can cause elevated AMR. The example includes MexAB-OprM efflux system in *P. aeruginosa*, which in presence of drugs is regulated positively. However, mutational changes in its regulator (*mexR*) results in overexpression of the MexAB-OprM and provide higher resistance to antibiotics for example, β-lactams [69].

Presently, five different classes of bacterial efflux machinery are reported [72], which includes: (i) ATP-binding cassette ABC, (ii) MATE-multidrug and toxic efflux, (iii) SMR-small MDR, (iv) RND-resistance nodulation division and (v) MF-major facilitator These families differ in their structure, function, energy source and the bacterial strain in which they are distributed.

7.3.3.2 Outer membrane permeability changes

The OM of gram −ve bacteria comprises of an inner phospholipid layer and an outer lipopolysaccharides (LPS) layer called lipid A. This OM complexity reduces the penetration of drug. The transport is attained through porin proteins present inside the OM and that forms water-filled channels. Antibiotics can invade the OM either by diffusion through (i) porin proteins or (ii) the lipid bilayer or by self-promoted uptake. The entry of antibiotic molecule is mainly dependent on its chemical composition. For example, antimicrobial agents which are β-lactams, fluoroquinolones and chloramphenicol penetrate the OM through porins [73, 74]. LPS is known to act as a barrier for the entry of antibiotics. Mutations in LPS structure have been reported to affect the antibiotic entry as well. The LPS defective mutated strains of *E. coli* and *S. enteric* serovar typhimurium are four times more susceptible to azithromycin, erythromycin and clarithromycin as compared to the wild-type strains [75, 76].

7.4 Antimicrobial susceptibility testing

Antimicrobial susceptibility testing (AST) is an in vitro practice used to determine the possibility of an antibiotic/antimicrobial agent in treating an infection caused by a specific pathogen/microorganism [77]. From the clinical microbiology aspects, the performance of AST is important to confirm the susceptibility to antibiotic/antimicrobial agents of choice, or to detect resistance in the individual pathogen/microorganisms [78]. The variety of experimental methods/protocols can be employed to determine the antimicrobial activity of an antibiotic/antimicrobial agent. However, the most basic methods of AST include the Kirby Bauer disk diffusion and broth or agar dilution methods [79]. Elucidation of results are particularly based on the Clinical Laboratory Standard Institute (CLSI) guidelines [80]. More recently, molecular methods namely polymerase chain reaction (PCR), DNA microarray, whole-genome sequencing, metagenomics and matrix-assisted laser desorption ionization–time of flight mass spectrometry (MALDI-TOF MS) have been introduced to detect and identify AMR genes [77, 80].

7.4.1 Disk diffusion/Kirby–Bauer method

Approved since 1975, this method uses a paper disk (6 mm size) impregnated with the desired concentration of antibiotic/antimicrobial agent, placed on a lawn of test bacteria [81]. The plates are incubated at 35 ± 2 °C for 24 h. After incubation, the inhibition zone is measured, and results are reported as resistant, susceptible and intermediate. The zone of inhibition arises when the amount of diffused antimicrobial/

antibiotic particles is high enough to inhibit microbial growth [77, 81]. To assess inhibitory zones, standardized readouts are available for many bacteria types and strains. The method is standardized to investigate certain fastidious bacterial pathogens such as *Streptococci, Neisseria gonorrhoeae, Haemophilus parainfluenzae, Haemophilus influenzae* and *Neisseria meningitides*; not for all fastidious bacteria [79]. The method is simple, low cost, easy to handle and capable of testing several microbes and antimicrobial agents [78]. However, unsuitability for slow and anaerobically growing microorganisms is a significant disadvantage of this method [81]. Furthermore, the method cannot distinguish between bacteriostatic and bactericidal effects because microbial growth inhibition does not necessarily mean the death of the microorganisms [79].

7.4.2 Agar well diffusion method

This method is primarily tested to estimate the antimicrobial potential of microbial or plant extracts. The method is similar to the disk diffusion method, where a volume of microbial inoculum containing 1–2×10^8 CFU/mL is spread on an agar plate. Then, the agar plate is aseptically punched with a sterilized tip or a cork borer to create a hole (6–8 mm diameter). The antimicrobial agent or extract (20–100 µL volume) is added to the well at the desired concentration followed by incubation of plates under appropriate conditions. The antimicrobial agent will diffuse into the agar plate and inhibit the growth of the test microorganisms [79].

7.4.3 Broth dilution method

This is one of the oldest and basic AST method for the estimation of MIC and minimum bactericidal concentration (MBC). The MIC is defined as the minimum amount of the tested antimicrobial compound that inhibits the growth of the microbial pathogen (test organism), and is generally expressed in µg/mL or mg/L. Whereas MBC is the lowest concentration of the drug at which 99.9% of bacteria are killed under standardized conditions given in document M26-A [79]. The broth dilution method has been optimized/standardized for aerobically growing bacteria, yeast and filamentous fungi by CLSI [82, 83].

The broth dilution method uses two-fold serial dilutions of antimicrobial agents (e.g., 2, 4, 8, 16 and 32 µg/mL) dispensed in the test tube having a minimum volume of 2 mL (macrodilution method). Whereas, in the microdilution method, smaller volume is dispensed in 96-well microtiter plate. In the subsequent step, a known concentration (5×10^5 CFU/mL as per CLSI recommendation) of suspended bacteria is inoculated and incubated for 24 h. After incubation, bacterial growth in the tubes is measured by turbidity [81]. In case of microdilution of fungi, the inoculum size defined by CLSI is the spore adjusted spectrophotometrically to 0.4×10^4–5×10^4 CFU/mL [79]. The MIC values can be influenced by inoculum size, incubation time, medium used and inoculum preparation method.

7.4.4 Agar dilution method

The method uses twofold dilution (e.g., 2, 4, 8, 16 and 32 µg/mL) of varying concentrations of antimicrobial agents incorporated into the agar medium, followed by the addition of defined concentration of microbial inoculum to be tested on agar plate surface [78]. The MIC endpoint is noted as the minimum amount of antimicrobial drug, which inhibits the growth of microbial inoculum added under suitable incubation conditions. The agar dilution method is favored over broth dilution in case if a single compound is being tested against multiple isolates, and the compounds/extract shows turbidity or coloration in the liquid medium. The method is recommended as standard for the microorganisms such as anaerobes and *Helicobacter* species [79].

7.4.5 E-test

The method is similar to disc diffusion method, which uses a predefined exponential gradient of antimicrobial agent applied to the bottom of a plastic strip instead of paper discs. The strip is then placed on agar medium plate having a defined microbial inoculum to allow the diffusion of drug. After 24 h of incubation, the MIC required to stop microbial growth is read on the strip. The test has advantages over the disc diffusion method as it provides exact MIC values, gradient stability for up to 20 h and suitable for numerous pathogens such as rapidly growing anaerobic and aerobic bacteria and slow growing fastidious bacteria [81, 84].

7.4.6 Future alternatives to AST

Over the last decade, several AST methods have been developed, but many of them are still far from routine clinical microbiology laboratory testing. Bacteriophage base-AST method has been well adopted for the testing of *M. tuberculosis* [85]. MALDI-TOF MS is an influential tool for the speedy identification of clinically important microorganisms. It may also prove to be of high importance as an AST method. Multiple methods have been investigated and documented, including (i) confirming the antibiotic-inactivating enzymes activity (e.g., -lactamases), (ii) presence of a PCR product indicative of AMR (e.g., mecA, NDM-1 or vanA) and (iii) proteomics analysis of the microorganisms in the presence or absence of an antimicrobial agent that correlates with susceptibility changes [86, 87].

The nucleic acid based AST methods have not been adopted extensively as they are inept to detect the entire resistance marker and are expensive. However, detection of resistant determinants by multiplex PCR directly from blood culture has been shown to attain timely and clinically actionable results. The nucleic acid diagnostics may be facilitated further by digital PCR and refinement of the aptamer technology [87].

The real-time microscopy is one among the innovative technologies that may be useful for AST in the very near future. The camera-based systems with better resolution have been commercialized and are proved to be highly effective. These techniques can produce a "1-day AST" for bacteria from positive blood culture bottles [87, 88]. Clearly, all these modern technologies offer near-future alternatives to resistance prediction as databases mature, and until then, the prospects of these technologies will remain nebulous.

7.5 Multidrug resistance

Before the introduction of chemotherapeutic agents, a large number of human deaths were caused due to microbial diseases. However, with the advent of antibiotics and other chemical therapeutic agents, the microbial diseases ceased to be a challenge for the medical community. But the extensive and often unplanned and unnecessary consumption of antibiotics has stemmed the development of multiple drug-resistant microorganisms [89]. These organisms are becoming as a grave concern for the world today.

The antibacterial agents have mainly five modes of action which are inhibition of cell wall biosynthesis (e.g., penicillins, beta-lactams, carbapenems and bacitracin), inhibition of protein synthesis (e.g., macrolides, aminoglycosides, tetracycline), disruption of membranes (e.g., polymyxins), inhibition of nucleic acid synthesis (like quinolones, rifampicin) and inhibition of folic acid synthesis (like sulfonamide). In case of MDR, a microorganism becomes resistant to various antimicrobial medicines having different structures and molecular targets, despite earlier sensitivity to them [90]. Because of this, the MDR microbes are able to survive the presence of antimicrobial drugs, resulting in ineffective treatment which in turn leads to persistence and spreading of infections. The MDRs, also sometimes referred to as "super bugs," cause prolonged illness, severe symptoms, higher morbidity and mortality rates. They pose a serious threat to the medical field apart from causing economic losses due to prolonged treatment regimes [91]. Some of the commonly encountered MDR microbes include methicillin-resistant *S. aureus* (MRSA), Vancomycin-resistant *Enterococci* (VRE), ESBL containing bacteria, XDR-TB (extensively drug-resistant TB) and so on. Many fungi like of *Candida* spp., *Aspergillus* spp., *Cryptococcus neoformans*, *Trichosporon beigeli* and so on have been reported to be resistant to macrolides, fluconazole, itraconazole, flucytosine, echinocandins and so on. [92]. The problem of drug resistance has also been encountered in many viruses like hepatitis B virus, influenza virus, cytomegalovirus, herpes simplex virus, *Varicella zoster* virus, HIV and so on, especially in case of immune-compromised patients [93]. Similar reports are also there about the development of resistance in many clinically important protozoan parasites like *Plasmodium, Leishmania, Entamoeba, Trichomonas vaginalis, Toxoplasma gondii* against commonly used drugs like chloroquine, pyrimethamine, artemisinin, antimonials, paromomycin, amphotericin B, sulfadiazine and so on. [94].

Microbes employ a variety of strategies for overcoming the ill effects of chemotherapeutic agents. Some microbes have been observed to alter the site of action of the drugs and thus rendering themselves resistant. This has been observed in case of many microbes which modify the topoisomerase enzymes (which serve as the target of fluoroquinolones) and hence are able to live even in the presence of the fluoroquinolones [95]. Some bacteria like *S. pneumoniae* and *Neisseria meningitides* have been reported to possess some mosaic PBPs which had lower affinity toward penicillin and hence leading to higher resistance against penicillins [96]. Also, the VREs have been observed to employ an altered precursor in cell wall biosynthesis, thus making vancomycin ineffective [97]. Some gram –ve bacteria have been reported to make use of efflux pump which actively pumps out tetracyclines from the cell while other microorganisms tend to prevent the entry of antimicrobial drugs [98]. Several microbes produce various enzymes which directly act on the drugs and lead to their inactivation. Such enzymes include aminoglycoside phosphoryltransferase, aminoglycoside acetyltransferase, aminoglycoside adenyltransferase, β-lactamases and so on [99]. Some of the significant multidrug-resistant microorganisms are discussed below.

7.5.1 Methicillin-resistant *Staphylococcus aureus*

S. aureus is a gram +ve, coccus-shaped, pyogenic bacterium (generally appearing in grape-bunch-like clusters), with capability to cause skin and cutaneous infections like cellulitis, abscesses, folliculitis, carbuncles, erysipelas, impetigo as well as severe and invasive conditions like pyomyositis, necrotizing fasciitis, osteomyelitis, necrotizing pneumonia and so on. [100]. The *S. aureus* strain which are resistant to the commonly used antibiotics are referred as methicillin-resistant *S. aureus*. Apart from methicillin, these bacteria may be resistant to other β-lactam antibiotics as well as other antibacterial agents [101].

Penicillin was introduced and became widely used in earlier parts of 1940s (especially during the WWII); however, within a short duration of time, the penicillin-resistant strains of *S. aureus* had emerged [102]. These bacteria produced a β-lactamase enzyme which hydrolyzed the critical β-lactam ring of the penicillin, thus rendering it harmless [103]. This enzyme is a serine β-lactamase encoded by blaZ gene, present in the transposon Tn*552* or similar elements [104] located on a large plasmid or is incorporated into the chromosome of bacteria [103]. In order to protect the antibiotic from bacterial enzyme, semisynthetic penicillins like methicillin and oxacillin were synthesized. Methicillin was introduced in 1959, and by 1961, MRSA had started appearing [104]. Within a short span, it spread across the globe and is now considered endemic to most healthcare institutions and is labeled as healthcare-associated MRSA (HA-MRSA). With time, MRSA has also been commonly reported from healthy members of human population and has been termed as community-associated MRSA (CA-MRSA) [105]. The MRSA infection is not limited to human beings but has also been reported in animals; such strains of MRSA are termed as livestock-associated MRSA (LA-MRSA) [106]. Reports of MRSA transmission between human and animals and vice versa are also there [107].

The resistance of MRSA to methicillin and oxacillin is mainly due to genes encoding a homologue of the PBP2 (which is the primary target of these antibiotics) called PBP2a or PBP2′ which is not susceptible to action of those antibacterial agents [108]. The active site serine of transpeptidase of PBP2a is not accessible to β-lactams since it is located deep inside a pocket [109]. PBP2a is encoded by the *mecA* gene which is sited within a family of discrete but related staphylococcal chromosome cassette (SCC) elements [110]. However, a distinct PBP2a encoded by mecC has also been reported [111].

In staphylococci, two Tet efflux pumps named TetA(K) and TetA(L), which exchange a proton for a tetracycline molecule against a concentration gradient have been reported [112]. The gene of TetK is located on a small multicopy plasmid pT181 and is integrated within the chromosomal SCC mecIII cassette of MRSA strains. These microbes have also been reported to carry tetO/M genes which are typically encoded on transposons like Tn*916* and Tn*1545* integrated into the chromosome [104]. The TetO/M protein binds to the EF-G binding site on the ribosome and dislodges bound tetracycline from the A site [113], hence allowing the microbe to carry on translation and surviving even in the presence of tetracyclines. Apart from this, some strains of MRSA have been reported to possess a *aacA-aphD* gene which encodes acetyltransferase–phosphotransferase which confers resistance to gentamcin and neomycin (by modifying them) and also a phosphotransferase (*aphA*) encoded by Tn*5405* leading to resistance against neomycin [114].

7.5.2 Vancomycin-resistant *Enterococci*

Enterococci are gram + ve, coccus-shaped bacteria which generally appear in chains. These are part of the normal intestinal microflora, however many of these bacteria can cause infections and diseases. Enterococci are capable of not only surviving in the gut of human beings and animals but can also survive in hospital environments and due to genetic adaptability can acquire resistance to various infections which are relatively difficult to treat [115]. VRE, belonging to the species *E. faecium*, was first observed in clinical isolates in Europe in 1986, followed by isolation of VR *E. faecalis* in the US in 1987 [116]. VRE possesses resistance against many types of antibiotics apart from being resistant to the relatively newer antibiotics like vancomycin.

Enterococci have a low level of inherent resistance to β-lactam compounds because of the fact that their PBPs show an intrinsic low affinity for them. However, many enterococci exhibit high-level β-lactam resistance. This phenomenon can be attributed to two possible reasons which are overproduction of low-affinity PBP5 and synthesis of β-lactamases. The overproduction of PBP5 with low-affinity binding to β-lactams makes the enterococci much less susceptible to these antibiotics and the β-lactamases tend to hydrolyze the β-lactam rings of these antibiotics thus protecting the cell from their harmful effect [117]. Also, the enterococci are known to be intrinsically resistant to low levels of aminoglycosides. This is because of lower penetrability of the enterococci toward the aminoglycosides [118]. The level resistance to aminoglycosides like gentamicin and streptomycin can be attributed to the production of aminoglycoside-modifying enzymes and alteration of ribosomal attachments sites [119].

Glycopeptidic antibacterial agents like vancomycin and teicoplanin target the biosynthesis of cell wall peptidoglycan by tightly binding with the d-Ala-d-Ala termini of pentapeptide precursors and thus preventing the cross-linking of peptidoglycan [97]. The enterococci overcome this problem by synthesizing alternate low-affinity pentapeptide precursors with d-Ala-d-Lac or d-Ala-d-Ser. The precursor d-Ala-d-Lac is associated with the phenotypes VanA, VanB, VanD, VanM, while the d-Ala-d-Ser precursor is associated with VanC, VanE, VanG, VanL, VanN [120]. The genetic elements encoding the resistance factors have been observed to be both plasmid encoded as well as chromosomally located [118].

7.5.3 Extensively drug-resistant TB

The tuberculosis is caused by *Mycobacterium tuberculosis* which is a non-spore forming, rod-shaped, acid-fast bacterium. The emergence of drug-resistant tuberculosis (DR-TB) closely followed the introduction of antitubercular chemotherapy, but the extensively drug-resistant TB (XDR-TB) is becoming a grave menace being faced by the world today [121]. As per WHO estimates, there were 170,000 DR-TB-related deaths in 2012 worldwide. The XDR-TB has been defined by the WHO as "TB with resistance to at least isoniazid and rifampicin as well as further resistance to any fluoroquinolones and second-line injectable drugs (SLIDs)." Another term XXDR-TB (or totally drug-resistant TB, i.e., TDR-TB) was proposed to indicate resistance to all available first-line drugs and second-line drugs (SLDs) [122]. Drug resistance in *M. tuberculosis* is because of the mutations which are favored due to the selective pressure resulting from the mismanagement of TB cases like improper treatment, not following proper treatment guidelines, inappropriate, incomplete or erratic use of SLDs and use of poor-quality SLDs [123]. The mutational changes in genes encoding drug-activating enzymes or drug targets lead to appearance of resistance in the bacteria [124].

M. tuberculosis treatment was conventionally done with the help of first line drugs like isoniazid and rifampicin. Isoniazid acts as a prodrug which is triggered by the enzyme catalase/peroxidase encoded by the katG gene. The triggered isoniazid binds with inhA gene product inhibiting mycolic acid synthesis [125]. The resistance to isoniazid is mediated by mutations in inhA, katG gene or their regulatory sequence [126]. Rifampicin binds to the RNA polymerase b subunit (encoded by gene rpoB), resulting in mRNA elongation inhibition. Rifampicin resistance is mediated by mutations clustered in codons 507–533 of rpoB. This altered b subunit is not bound by high affinity to rifampicin thus overcoming the inhibitory effect of rifampicin [127].

Pyrazinamide, another first line drug, is triggered by the enzyme pyrazinamidase/ nicotinamidase (PZase) to pyrazinoic acid. The activating enzyme is encoded by the pncA gene [128]. The pyrazinoic acid disintegrates the bacterial membrane energetics leading to inhibition of membrane transport. It has also been observed to interfere with fatty acid synthesis [129]. Mutations in the pncA gene and its promoter region which prevent conversion of the prodrug into active drug are the most common mechanism mediating pyrazinamide resistance [130].

Another drug ethambutol interferes with biosynthesis of the arabinogalactan (a component of cell wall) by targeting arabinosyl transferase enzyme (encoded by embB gene). Mycobacteria become resistant to it by via mutations in the embB gene with modification in codon 306 being the most common change leading to resistance [131]. Similarly in case of Streptomycin the main mechanism of resistance is via mutations in rpsL and rrs genes, encoding the ribosomal protein S12 and16S rRNA which are the targets of streptomycin. [132].

The SLIDs or agents include aminoglycosides like kanamycin, amikacin and the cyclic polypeptide capreomycin. All of these act as protein biosynthesis inhibitors by binding to ribosomes resulting in alteration of 16S rRNA structure. Bacterial resistance to them have been mostly attributed to mutations in the 1,400 bp region of the rrs gene and polymorphisms of the tlyA gene [133]. Fluoroquinolones that target DNA gyrase enzymes, encoded by genes gyrA and gyrB are effective bactericidal antibiotics presently used as second-line treatment. Most of the resistance to the fluoroquinolones arise due to mutations occurring in a conserved, quinolone resistance-determining region in gyrA and gyrB genes [134].

Mycobacteria have also evolved resistance to many novel drugs. Bedaquiline is a new drug that belongs to a novel class of diarylquinolines, which has been known to target mycobacterial ATP synthase, inhibiting bacterial respiration. Mutations in the atpE gene (encoding F1F0 proton ATP synthase) have been associated with resistance to bedaquiline [135]. Another drug linezolid binds to 50S ribosomal subunit, thereby interfering in protein synthesis. Resistance to linezolid has been associated with mutations in the 23S rRNA gene [136]. Apart from this, efflux pump systems are involved in development of resistance. *M. tuberculosis* possesses very high number of possible efflux pumps with 148 genes coding membrane transport proteins. The overexpression of efflux pumps is supposed to be responsible for increased level of resistance mutations to a variety of drugs [137].

7.5.4 Drug-resistant viruses

The problem of drug resistance is not confined to bacteria as there are several clinically important viruses which have been reported to display resistance against many antiviral drugs [93]. High mutation rate of viruses is especially significant in the emergence of antiviral drug resistance. The influenza A virus is a RNA virus containing a segmented genome. One of the major drugs used for its control is oseltamivir. However, the mutation known as H274Y reported in the virus was reported to confer very high levels of resistance to osteltamivir [138]. Being a DNA virus, the herpes simplex virus shows relatively low mutation rate as compared to the RNA viruses [139]. The most common drug used for its control is the Acyclovir which acts as a nucleoside analog and there are already reports of viral mutants which impact thymidine kinase or viral DNA polymerase, thus rendering the mutant resistant to the drug [140]. Similarly, in case of cytomegalovirus, the resistant mutants were observed in the viral DNA polymerase, thus making the virus resistant to nucleoside analog drugs like ganciclovir and cidofovir [141].

7.6 Conclusions

The use, misuse and abuse of antibiotics have emerged as the most pressing healthcare challenge of the new millennium as this may lead to antibiotic resistance. The environmental bacteria were thought to be the cause of AMR as clinically relevant resistance genes were found on the chromosomes of these bacteria. The literature acquaints that the antibiotic resistance is proliferating at an alarming rate. The microbial infections are becoming harder/impossible to treat and antibiotics are becoming ineffective. Meanwhile, the development of novel antibiotic compounds have decreased over the decades. Moreover, there is very limited antibiotic treatment repertoire available for existing and emerging MDR bacteria, resulting in high mortality and morbidity. Extensive efforts are required to reduce the use of antibiotics and to find/discover other alternate treatment strategies. The pace of drug resistance can be minimized by studying and exploring new antimicrobial agents, resistance mechanisms and emerging microbes. The chapter covers introduction to AMR, biochemical and molecular aspects of resistance and methods of susceptibility testing and MDR.

References

[1]   Baker SJ, Payne DJ, Rappuoli R, De Gregorio E, Technologies to address antimicrobial resistance, Proc Natl Acad Sci USA, 2018, 115(51), 12887–95.

[2]   Kaur P, Peterson E, Antibiotic resistance mechanisms in bacteria: Relationships between resistance determinants of antibiotic producers, environmental bacteria, and clinical pathogens, Front Microbiol, 2018, 9, 2928.

[3]   Founou RC, Founou LL, Essack SY, Clinical and economic impact of antibiotic resistance in developing countries: A systematic review and meta-analysis, PloS One, 2017, 12(12), e0189621.

[4]   O'neill JI, Antimicrobial resistance: Tackling a crisis for the health and wealth of nations, Rev Antimicrob Resist, 2014, 20, 1–6.

[5]   Laura VM, Vladi MH, Vasile SC Determining the Antibiotic Resistance of Bacterial Pathogens in Sexually Transmitted Diseases. Antibacterial Agents. 2017:109.

[6]   Cheesman MJ, Ilanko A, Blonk B, Cock IE, Developing new antimicrobial therapies: Are synergistic combinations of plant extracts/compounds with conventional antibiotics the solution?, Phcog Rev, 2017, 11(22), 57.

[7]   Castro-Sánchez E, Moore LS, Husson F, Holmes AH, What are the factors driving antimicrobial resistance? Perspectives from a public event in London, England, BMC Infect Dis, 2016, 16(1), 465.

[8]   Ghosh C, Sarkar P, Issa R, Haldar J, Alternatives to conventional antibiotics in the era of antimicrobial resistance, Trends Microbiol, 2019, https://doi.org/10.1016/j.tim.2018.12.010.

[9]   Howard SJ, Arendrup MC, Acquired antifungal drug resistance in *Aspergillus fumigatus*: Epidemiology and detection, Med Mycol, 2011, 49, S90–5.

[10] Hurt AC, The epidemiology and spread of drug resistant human influenza viruses, Curr Opin Virol, 2014, 8, 22–29.

[11] Loeffler J, Stevens DA, Antifungal drug resistance, Clin Infect Dis, 2003, 36, S31–41.

[12] Lurain NS, Chou S, Antiviral drug resistance of human cytomegalovirus, Clin Microbiol Rev, 2010, 23(4), 689–712.

[13] Tacconelli E, Carrara E, Savoldi A, Harbarth S, Mendelson M, Monnet DL, Pulcini C, Kahlmeter G, Kluytmans J, Carmeli Y, Ouellette M, Discovery, research, and development of new antibiotics: The WHO priority list of antibiotic-resistant bacteria and tuberculosis, Lancet Infect Dis, 2018, 18(3), 318–27.

[14] Tanwar J, Das S, Fatima Z, Hameed S, Multidrug resistance: An emerging crisis, Interdiscip Perspect Infect Dis, 2014, 2014.

[15] Ventola CL, The antibiotic resistance crisis: Part 1: Causes and threats, P & T, 2015, 40(4), 277.

[16] Frost LS, Leplae R, Summers AO, Toussaint A, Mobile genetic elements: The agents of open source evolution, Nat Rev Microbiol, 2005, 3(9), 722–32.

[17] Dhar S, Kumari H, Balasubramanian D, Mathee K, Cell-wall recycling and synthesis in Escherichia coli and *Pseudomonas aeruginosa* – their role in the development of resistance, J Med Microbiol, 2018, 67, 1–21.

[18] Rumbo C, Gato E, Lopez M, et al., Contribution of efflux pumps, porins, and beta-lactamases to multidrug resistance in clinical isolates of *Acinetobacter baumannii*, Antimicrob Agents Chemother, 2013, 57, 5247–57.

[19] Beceiro A, Tomas M, Bou G, Antimicrobial resistance and virulence: A beneficial relationship for the microbial world?, EnfermInfecc Microbiol Clin, 2012, 30, 492–99.

[20] Harbottle H, Thakur S, Zhao S, White DG, Genetics of antimicrobial resistance, AnimBiotechnol, 2006, 17, 111–24.

[21] Alekshun MN, Levy SB, The mar regulon: Multiple resistance to antibiotics and other toxic chemicals, Trends Microbiol, 1999, 7, 410–13.

[22] Hricova K, Kolar M, Bacterial efflux pumps – Their role in antibiotic resistance and potential inhibitors, Klin MikrobiolInfekc Lek, 2014, 20, 116–20.

[23] Correia S, Poeta P, Hebraud M, Capelo JL, Igrejas G, Mechanisms of quinolone action and resistance: Where do we stand?, J Med Microbiol, 2017, 66, 551–59.

[24] Grigorenko VG, Rubtsova MY, Uporov IV, Ishtubaev IV, Andreeva IP, Shcherbinin DS, Veselovsky AV, Egorov AM, Bacterial TEM-type serine beta-lactamases: Structure and analysis of mutations, Biomed Khim, 2017, 63, 499–507.

[25] Bush K, Pa B, beta-Lactams and beta-Lactamase Inhibitors: An Overview, Cold Spring Harb Perspect Med, 2016, Vol. 6.

[26] Partridge SR, Kwong SM, Firth N, Jensen SO, Mobile genetic elements associated with antimicrobial resistance, Clin Microbiol Rev, 2018, 31(4), e00088–17.

[27] Li Q, Chang W, Zhang H, Hu D, Wang X, The Role of Plasmids in the Multiple Antibiotic Resistance Transfer in ESBLs-Producing *Escherichia coli* Isolated From Wastewater Treatment Plants, Front Microbiol, 2019, 10, 633.

[28] Andres P, Lucero C, Soler-Bistué A, Guerriero L, Albornoz E, Tran T, Zorreguieta A, Galas M, Corso A, Tolmasky ME, Petroni A, Differential distribution of plasmid-mediated quinolone resistance genes in clinical enterobacteria with unusual phenotypes of quinolone susceptibility from Argentina, Antimicrob Agents Chemother, 2013, 57(6), 2467–75.

[29] Recchia GD, Hall RM, Origins of the mobile gene cassettes found in integrons, Trends Microbiol, 1997, 5(10), 389–94.

[30] Sultan I, Rahman S, Jan AT, Siddiqui MT, Mondal AH, Haq QM, Antibiotics, resistome and resistance mechanisms: A bacterial perspective, Front Microbiol, 2018, 9.

[31] Johnson CM, Grossman AD, Integrative and Conjugative Elements (ICEs): What They Do and How They Work, Annu Rev Genet, 2015, 49, 577–601.

[32] Deng Y, Bao X, Ji L, Chen L, Liu J, Miao J, Chen D, Bian H, Li Y, Yu G, Resistance integrons: Class 1, 2 and 3 integrons, Ann Clin Microbiol Antimicrob, 2015, 14, 45.

[33] Mobashery S, Azucena EF, Bacterial Antibiotic Resistance, Encyclopedia of Life Sciences, London, UK, Nature Publishing Group, 1999.

[34] Wright GD, Bacterial resistance to antibiotics: Enzymatic degradation and modification, Adv Drug Deliv Rev, 2005, 57, 1451–70.

[35] Pa L, Adv Drug Deliv Rev, 2005, 57, 1471–85.

[36] Kumar A, Schweizer HP, Bacterial resistance to antibiotics: Active efflux and reduced uptake, Adv Drug Deliv Rev, 2005, 57, 1486–513.

[37] Happi CT, Gbotosho GO, Folarin OA, Akinboye DO, Yusuf BO, Ebong OO, Sowunmi A, Kyle DE, Milhous W, Wirth DT, Oduola AMJ, Polymorphisms in *Plasmodium falciparum* dhfr and dhps genes and age related invivosulfaxine-pyrimethamine resistance in malaria-infected patients from Nigeria, Acta Trop, 2005, 95, 183–93.

[38] Kaur P, Peterson E, Antibiotic resistance mechanisms in bacteria: Relationships between resistance determinants of antibiotic producers, environmental bacteria, and clinical pathogens, Front Microbiol, 2018, 9, 2928.

[39] Poole K, Resistance to β-lactam antibiotics, Cell Mol Life Sci, 2004, 61, 2200–23.

[40] Bonnet R, Growing group of extended-spectrum beta-lactamases: The CTX-M enzymes, Antimicrob Agents Chemother, 2004, 48, 1–14.

[41] Bradford PA, Extended-spectrum β-lactamases in the 21[st]century: Characterization, epidemiology, and detection of this important resistance threat, Clin Microbiol Rev, 2001, 14, 933–51.

[42] Shah AA, Hasan F, Ahmed S, Hameed A, Extended–spectrum β-lactamases (ESBLs): Characterization, epidemiology and detection, Crit Rev Microbiol, 2004, 30, 25–32.

[43] Vetting MW, Magnet S, Nieves E, Roderick SL, Blanchard JS, A bacterial acetyltransferase capable of regioselective N-acetylation of antibiotics and histones, Chem Biol, 2004, 11, 565–73.

[44] Schwarz S, Kehrenberg C, Doublet B, Cloeckaert A, Molecular basis of bacterial resistance to chloramphenicol and florfenicol, FEMS Microbiol Rev, 2004, 28, 519–42.

[45] Nakamura A, Miyakozawa I, Nakazawa K, Sawai OHKT, Detection and characterization of a macrolide 2′-phosphotransferasefrom *Pseudomonas aeruginosa* clinical isolate, Antimicrob Agents Chemother, 2000, 44, 3241–42.

[46] Matsuoka M, Sasaki T, Inactivation of macrolides by producersand pathogens, Curr Drug Targets Infect Disord, 2004, 4, 217–40.

[47] Houang AT, Chu YW, Lo WS, Chu KY, Cheng AF, Epidemiology of rifampin ADP-ribosyltransferase (arr-2) andmetallo-β-lactamase (blaIMP-4) gene cassettes in class 1 integronsin*Acinetobacter* strains isolated from blood culturesin 1997 to 2000, Antimicrob Agents Chemother, 2003, 47, 1382–90.

[48] Yang W, Moore IF, Koteva KP, Bareich DC, Hughes DW, Wright GD, TetX is a flavin-dependent monooxygenase conferring resistance to tetracycline antibiotics, J BiolChem, 2004, 279, 52346–52.

[49] Guengerich FP, Common and uncommon cytochrome P450reactions related to metabolism and chemical toxicity, ChemResToxicol, 2001, 14, 611–50.

[50] Spratt BG, Resistance to antibiotics mediated by target alterations, Science, 1994, 264, 388–93.

[51] Enright MC, The evolution of a resistant pathogen – Thecase of MRSA, Curr Opin Pharmacol, 2003, 3, 474–79.

[52] Leski TA, Tomasz A, Role of penicillin-binding protein 2(PBP2) in the antibiotic susceptibility and cell wall cross-linking of *Staphylococcus aureus*: Evidence for the cooperative functioning of PBP2, PBP4, and PBP2A, J Bacteriol, 2005, 187, 1815–24.

[53] Nagai K, Davies TA, Jacobs MR, Appelbaum PC, Effects of amino acid alterations in penicillin-binding proteins(PBPs) 1a, 2b, and 2x on PBP affinities of penicillin, ampicillin, amoxicillin, cefditoren, cefuroxime, cefprozil, and cefaclor in 18 clinical isolates of penicillin-susceptible, -intermediate, and -resistant pneumococci, Antimicrob Agents Chemother, 2002, 46, 1273–80.

[54] Kosowska K, Jacobs MR, Bajaksouzian S, Koeth L, Appelbaum PC, Alterations of penicillin-binding proteins 1A,2X, and 2B in *Streptococcus pneumoniae* isolates for which amoxicillin MICs are higher than penicillin MICs, Antimicrob Agents Chemother, 2004, 48(10), 4020–22.

[55] Hiramatsu K, Vancomycin-resistant *Staphylococcus aureus*: A new model of antibiotic resistance, Lancet Infect Dis, 2001, 1, 147–55.

[56] Cooper MA, Fiorini MT, Abell C, Williams DH, Bindingof vancomycin group antibiotics to D-alanine and D-lactatepresenting self-assembled monolayers, Bioorg Med Chem, 2000, 8, 2609–16.

[57] Spigaglia P, Mastrantonio P, Analysis of macrolide-lincosamide-streptogramin B (MLSB) resistance determinant instrains of *Clostridium difficile*, Microb Drug Resist, 2002, 8, 45–53.

[58] Ackermann G, Degner A, Cohen SH, Silva J Jr, Rodloff AC, Prevalence and association of macrolide-lincosamide-streptogramin B (MLSB) resistance with resistance tomoxifloxacin in *Clostridium difficile*, J Antimicrob Chemother, 2003, 51, 599–603.

[59] Weisblum B, Erythromycin resistance by ribosome modification, Antimicrob Agents Chemother, 1995, 39, 577–85.

[60] Wang G, Taylor DE, Site-specific mutations in the 23SrRNA gene of Helicobacter pylori confer two types of resistance to macrolide–lincosamide–streptogramin B antibiotics, Antimicrob Agents Chemother, 1998, 42, 1952–58.

[61] Suzuki Y, Katsukawa C, Tamaru A, Abe C, Makino M, Mizuguchi Y, Taniguchi H, Detection of kanamycin-resistant *Mycobacterium tuberculosis* by identifying mutationsin the 16S rRNA gene, J Clin Microbiol, 1998, 36, 1220–25.

[62] Maravic Vlahovicek G, Cubrilo S, Tkaczuk KL, Bujnicki JM, Modeling and experimental analyses reveal a two–domain structure and amino acids important for the activityof aminoglycoside resistance methyltransferase Sgm, Biochem Biophys Acta, 2008, 1784(4), 582–90.

[63] Khodursky AB, Zechiedrich EL, Cozzarelli NR, Topoisomerase IV is a target of quinolones in *Escherichia coli*, Proc Natl Acad Sci USA, 1995, 92, 11801–05.

[64] Ince D, Zhang X, Silver LC, Hooper DC, Dual targetingof DNA gyrase and topoisomerase IV: Target interactionsof garenoxacin (BMS-284756, T-3811ME), a new desfluoroquinolone, Antimicrob Agents Chemother, 2002, 46, 3370–80.

[65] Eliopoulos GM, Quinolone resistance mechanisms in pneumococci, Clin Infect Dis (Suppl), 2004, 38, 350–56.

[66] Nikaido H, Prevention of drug access to bacterial targets: Permeability barriers and active efflux, Science, 1994, 264, 382–88.

[67] Webber MA, Piddock LJ, The importance of efflux pumpsin bacterial antibiotic resistance, J Antimicrob Chemother, 2003, 51, 9–11.

[68] Van Veen HW, Konings WN, Drug efflux proteins in multidrugresistant bacteria, Biol Chem, 1997, 378, 769–77.

[69] Poole K, Multidrug efflux pumps and antimicrobial resistance in *P. aeruginosa* and related organisms, J Mol Microbiol Biotechnol, 2001, 3, 225–64.

[70] Langton KP, Henderson PJF, Herbert RB, Antibiotic resistance: Multidrug efflux proteins, a common transport mechanism?, Nat Prod Rep, 2005, 22, 439–51.

[71] Putman M, Van Veen HW, Konings WN, Molecular propertiesof bacterial multidrug transporters, Microbiol Mol Biol Rev, 2000, 64, 672–93.

[72] Pao SS, Paulsen IT, Saier MH Jr, Major facilitator superfamily, Microbiol Mol Biol Rev, 1998, 62, 1–34.

[73] Nikaido H, Molecular basis of bacterial outer membrane permeability revisited, Microbiol Mol Biol Rev, 2003, 67, 593–656.

[74] Denyer SP, Maillard JY, Cellular impermeability and uptake of biocides and antibiotics in Gram-negative bacteria, J Appl Microbiol (Suppl), 2002, 92, 35–45.

[75] Wiese A, Brandenburg K, Ulmer AJ, Seydel U, Müller-Loennies S, The dual role of lipopolysaccharide as effector and target molecule, Biol Chem, 1999, 380, 767–84.

[76] Stokes JM, French S, Ovchinnikova OG, Bouwman C, Whitfield C, Brown ED, Cold stress makes *Escherichia coli* susceptible to glycopeptide antibiotics by altering outer membrane integrity, Cell Cheml Boil, 2016, 23(2), 267–77.

[77] Wanger A, Chavez V, Huang RS, Wahed A, Actor JK, Dasgupta A, Antibiotics, Antimicrobial Resistance, Antibiotic Susceptibility Testing, and Therapeutic Drug Monitoring for Selected Drugs, Microbiol Mol Diagnosis Pathol, 2017, 119–53.

[78] Reller LB, Weinstein M, Jorgensen JH, Ferraro MJ, Antimicrobial susceptibility testing: A review of general principles and contemporary practices, Clin Infect Dis, 2009, 49(11), 1749–55.

[79] Balouiri M, Sadiki M, Ibnsouda SK, Methods for in vitro evaluating antimicrobial activity: A review, J Pharm Anal, 2016, 6(2), 71–79.

[80] Clinical and Laboratory Standards Institute (CLSI)., Performance Standards for Antimicrobial Susceptibility Testing, CLSI supplements M100S, 26th ed, 2016.

[81] Schumacher A, Vranken T, Malhotra A, Arts JJ, Habibovic P, In vitro antimicrobial susceptibility testing methods: Agar dilution to 3D tissue-engineered models, Eur J Clin Microbiol Infect Dis, 2018, 37(2), 187–208.

[82] Clinical and Laboratory Standards Institute., Methods for Dilution Antimicrobial Susceptibility Tests for Bacteria that Grow Aerobically, Approved Standard, 9th ed., CLSI document, 2012, M07-A9.

[83] Clinical and Laboratory Standards Institute., Reference Method for Broth Dilution Antifungal Susceptibility Testing Filamentous Fungi, Approved Standard, 2nd ed., CLSI document, 2008, M38-A2.

[84] Doddangoudar VC, O'Donoghue MM, Boost MV, Tsang DN, Appelbaum PC, Rapid detection of vancomycin-non-susceptible Staphylococcus aureus using the spiral gradient endpoint technique, J Antimicrob Chemother, 2010, 65(11), 2368–72.

[85] Wilson SM, Al-Suwaidi Z, McNerney R, Porter J, Drobniewski F, Evaluation of a new rapid bacteriophage-based method for the drug susceptibility testing of *Mycobacterium tuberculosis*, Nat Med, 1997, 3(4), 465–68.

[86] Van Belkum A, Welker M, Erhard M, Chatellier S, Biomedical mass spectrometry in today's and tomorrow's clinical microbiology laboratories, J Clinmicrobiol, 2012, 50(5), 1513–17.

[87] Van Belkum A, Dunne WM, Next-generation antimicrobial susceptibility testing, J Clinmicrobiol, 2013, 51(7), 2018–24.

[88] Douglas IS, Price CS, Overdier K, Thompson K, Wolken B, Metzger S, Howson D, Rapid microbiological identification and major drug resistance phenotyping with novel multiplexed automated digital microscopy (MADM) for ventilator-associated pneumonia (VAP) surveillance, Am J Respir Crit Care Med, 2011, 183, A3928.

[89] Rahman S, Ali T, Ali I, Khan NA, Han B, Gao J, The Growing Genetic and Functional Diversity of Extended Spectrum Beta-Lactamases, Bio Med Res Inter, 2018, Article ID 9519718, 1–15, https://doi.org/10.1155/2018/9519718.

[90] Singh V, Antimicrobial resistance, Microbial Pathogens and Strategies for Combating Them: Science, Technology and Education, Vol. 1, Badajoz, Spain, Formatex Research Center, 2013, 291–96.

[91] Tanwar J, Das S, Fatima Z, Hameed S, Multidrug Resistance: An Emerging Crisis, Interdis Pers Infec Dis, 2014, 1–8, https://doi.org/10.1155/2014/541340.

[92] Loeffler J, Stevens DA, Antifungal drug resistance, ClinInfec Dis, 2003, 36, S31–S41.

[93] Irwin KK, Renzette N, Kowalik TF, Jensen JD, Antiviral drug resistance as an adaptive process, Virus Evo, 2016, 2, 1–10.

[94] Mohapatra S, Drug resistance in leishmaniasis: Newer developments, Trop Parasitol, 2014, 4, 4–9.

[95] Hooper DC, Mechanisms of action and resistance of older and newer fluoroquinolones, Clin Infect Dis, 2000, 31, S24–28.

[96] Spratt BG, Resistance to antibiotics mediated by target alterations, Science, 1994, 264, 388–93.

[97] Courvalin P, Vancomycin resistance in gram-positive cocci, Clin Infect Dis, 2006, 42, S25–34.

[98] Chethana GS, Hari Venkatesh KR, Mirzaei F, Gopinath SM, Multidrug resistant bacteria and its implication in medical sciences, J Biol Sci Opin, 2013, 1, 32–37.

[99] Barlow M, Reik RA, Jacobs SD, Medina M, Meyer MP, et al., High rate of mobilization for *bla*CTXMs, Emerg Infect Dis, 2008, 14, 423–28.

[100] Green BN, Johnson CD, Egan JT, Rosenthal M, Griffith EA, Evans MW, Methicillin-resistant *Staphylococcus aureus*: An overview for manual therapists, J Chiroprac Med, 2012, 11, 64–76.

[101] Chambers HF, The changing epidemiology of *Staphylococcusaureus*?, Emerg Infect Dis, 2001, 7, 178–82.

[102] Walsh CT, Wencewicz TA, Antibiotics: Challenges, Mechanisms, Opportunities, Washington, DC, ASM Press, 2016.

[103] Lowy FD, Antimicrobial resistance: The example of *Staphylococcus aureus*, J Clin Invest, 2003, 111, 1265–73.

[104] Jensen SO, Lyon BR, Genetics of antimicrobial resistance in *Staphylococcus aureus*, Future Microbiol, 2009, 4, 565–82.

[105] Liu C, Bayer A, Cosgrove SE, et al., Clinical practice guidelines by the infectious diseases society of america for the treatment of methicillin-resistant *Staphylococcus aureus* infections in adults and children, Clin Infect Dis, 2011, 52, e18–55.

[106] Paterson GK, Larsen AR, Robb A, Edwards GE, Pennycott TW, Foster G, The newly described mecA homologue, mecALGA251, is present in methicillin-resistant *Staphylococcus aureus* isolates from a diverse range of host species, J Antimicrob Chemother, 2012, 67, 2809–13.

[107] Grema HA, Geidam YA, Gadzama GB, Ameh JA, Suleiman A, Methicillin resistant *Staphyloccus aureus* (MRSA): A review, Adv Anim Vet Sci, 2015, 3, 79–98.

[108] Fisher JF, Mobashery S, beta-lactam resistance mechanisms: Gram-positive bacteria and *Mycobacterium tuberculosis*, Cold Spring Harb Perspect Med, 2016, 6, a025221.

[109] Lim D, Strynadka NC, Structural basis for the beta lactam resistance of PBP2a from methicillin-resistant *Staphylococcus aureus*, Nat Struct Biol, 2002, 9, 870–76.

[110] Liu J, Chen D, Peters BM, Li L, *et al.*, Staphylococcal chromosomal cassettes mec (SCCmec): A mobile genetic element in methicillin-resistant *Staphylococcus aureus*, Microb Pathog, 2016, 101, 56–67.

[111] Paterson GK, Harrison EM, Holmes MA, The emergence of mecC methicillin-resistant *Staphylococcus aureus*, Trends Microbiol, 2014, 22, 42–47.

[112] Piddock LJ, Multidrug-resistance efflux pumps – Not just for resistance, Nat Rev Microbiol, 2006, 4, 629–36.

[113] Connell SR, Trieber CA, Dinos GP, et al., Mechanism of Tet(O)-mediated tetracycline resistance, EMBO J, 2003, 22, 945–53.

[114] Foster TJ, Antibiotic resistance in *Staphylococcus aureus*. Current status and future prospects, FEMS Microbiol Rev, 2017, 41, 430–49.

[115] Ahmed MO, Baptiste KE, Vancomycin-Resistant Enterococci: A Review of Antimicrobial Resistance Mechanisms and Perspectives of Human and Animal Health, Microb Drug Resist, 2018, 24, 590–606.

[116] Uttley AH, Collins CH, Naidoo J, George RC, Vancomycin-resistant enterococci, Lancet, 1988, 1, 57–58.

[117] Arias CA, Contreras GA, Murray BE, Management of multidrug-resistant enterococcal infections, Clin Microbiol Infect, 2010, 16, 555–62.

[118] O'Driscoll T, Crank CW, Vancomycin-resistant enterococcal infections: Epidemiology, clinical manifestations, and optimal management, Infec Drug Resist, 2015, 8, 217–30.

[119] Mainardi JL, Villet R, Bugg TD, Mayer C, Arthur M, Evolution of peptidoglycan biosynthesis under the selective pressure of antibiotics in Gram-positive bacteria, FEMS Microbiol Rev, 2008, 32, 386–408.

[120] Lebreton F, Depardieu F, Bourdon N, et al., D-Ala-d-SerVanN-type transferable vancomycin resistance in *Enterococcus faecium*, Antimicrob Agents Chemother, 2011, 55, 4606–12.

[121] Shah NS, Wright A, Drobniewski F, Extreme drug resistance in tuberculosis (XDR-TB): Global survey of supranational reference laboratories for *Mycobacterium tuberculosis* with resistance to second-line drugs, Int J Tuberc Lung Dis, 2005, 9, S77.

[122] Migliori GB, Loddenkemper R, Blasi F, Raviglione MC, 125 years after Robert Koch's discovery of the tubercle bacillus: The new XDR-TB threat. Is "science" enough to tackle the epidemic?, Eur Respir J, 2007, 29, 423–27.

[123] Prasad R, Singh A, Balasubramanian V, Gupta N, Extensively drug-resistant tuberculosis in India: Current evidence on diagnosis & management, Indian J Med Res, 2017, 145(3), 271–93.

[124] Günther G, Multidrug-resistant and extensively drug-resistant tuberculosis: A review of current concepts and future challenges, Clin Med, 2014, 14, 279–85.

[125] Palomino JC, Martin A, Drug resistance mechanisms in *Mycobacterium tuberculosis*, Antibiotics (Basel), 2014, 3, 317–40.

[126] Vilche`ze C, Jacobs J, William R, Themechanismof isoniazid killing: Clarity through the scope of genetics, Annu Rev Microbiol, 2007, 61, 35–50.

[127] Caws M, Duy PM, Tho DQ etal.Mutations prevalent among rifampin- and isoniazid-resistant *Mycobacteriumtuberculosis* isolates from a hospital in Vietnam, J Clin Microbiol, 2006, 44, 2333–37.

[128] Zhang Y, Mitchison D, The curious characteristics of pyrazinamide: A review, Int J Tuberc Lung Dis, 2003, 7, 6–21.

[129] Zimhony O, Vilche`ze C, Arai M, et al., Pyrazinoic acid and its n-propyl ester inhibit fatty acid synthase type I in replicating tubercle bacilli, Antimicrob Agents Chemother, 2007, 51, 752–54.

[130] Pandey B, Grover S, Tyagi C, et al., Molecular principles behind pyrazinamide resistance due tomutations in panD gene in *Mycobacterium tuberculosis*, Gene, 2016, 581, 31–34.

[131] Safi H, Lingaraju S, Amin A, et al., Evolution of high-level ethambutol resistant tuberculosis through interacting mutations in decaprenylphosphoryl-b-D-arabinose biosynthetic and utilization pathway genes, Nat Genet, 2013, 45, 1190–97.

[132] Spies FS, Da Silva PEA, Ribeiro MO, et al., Identification of mutations related to streptomycin resistance in clinical isolates of *Mycobacterium tuberculosis* and possible involvement of efflux mechanism, Antimicrob Agents Chemother, 2008, 52, 2947–49.

[133] Reeves AZ, Campbell PJ, Willby MJ, et al., Disparities in capreomycin resistance levels associated with the rrs A1401G mutation in clinical isolates of *Mycobacterium tuberculosis*, Antimicrob Agents Chemother, 2015, 59, 444–49.

[134] Maruri F, Sterling TR, Kaiga AW, et al., A systematic review of gyrase mutations associated with fluoroquinolone-resistant *Mycobacterium tuberculosis* and a proposed gyrase numbering system, J Antimicrob Chemother, 2012, 67, 819–31.

[135] Pontali E, Sotgiu G, D'Ambrosio L, et al., Bedaquiline and multidrug resistant tuberculosis: A systematic and critical analysis of the evidence, Eur Respir J, 2016, 47, 394–402.

[136] Dookie N, Rambaran S, Padayatchi N, Mahomed S, Naidoo K, Evolution of drug resistance in *Mycobacterium tuberculosis*: A review on the molecular determinants of resistance and implications for personalized care, J Antimicrob Chemother, 2018, 73, 1138–51.

[137] Machado D, Couto I, Perdig~ao J, et al., Contribution of efflux to the emergence of isoniazid and multidrug resistance in *Mycobacterium tuberculosis*, PLoS One, 2012, 7, e34538.

[138] Moscona A, Global Transmission of Oseltamivir-Resistant Influenza, New Eng J Med, 2009, 360, 953–56.

[139] Andrei G, Snoeck R, Herpes Simplex Virus drug-resistance: New mutations and insights, Curr OpinInfec Dis, 2013, 26, 551–60.

[140] Griffiths A, Slipping and Sliding: Frameshift Mutations in Herpes Simplex Virus Thymidine Kinase and Drug- Resistance, Drug Resist Updates, 2011, 14, 251–59.

[141] Gilbert C, Boivin G, Human Cytomegalovirus Resistance to Antiviral Drugs, Antimicrob Agents Chemother, 2005, 49, 873–83.

Nazia Tarannum, Ranjit Hawaldar

Chapter 8
Microbiology as an occupational hazard: risk and challenges

Abstract: A large number of cases have been reported worldwide regarding occupationally acquired microbial diseases. These cases are of major concern because they have potential to cause pandemics. Among the various occupational health hazards, the most serious one is the risk caused by biological agents like viruses, fungi, bacteria and human parasites. The direct/indirect contact with animals, infectious materials and cultures also pose a health hazard. In current chapter, role of microbiology hazards in various occupations, occupational zoonotic diseases, legislation, safety policies and biosafety measures in microbiological laboratories have been discussed.

Keywords: microbial disease, health hazard, infection, biosafety, policies

8.1 Components of microbiological wastes: introduction

Among the various occupational health hazards, the most serious one is the risk caused by biological agents like viruses, fungi, bacteria and human parasites and direct/indirect contact with animals, infectious materials and cultures. There are several factors, which decide the biological agent's pathogenicity and entry route of pathogens into the human body. For a professional setting, susceptibility of workers, the dose needed to cause infection and the transmission mode are important [1–3]. Thousands of cases have been reported regarding occupationally acquired hepatitis B, human immunodeficiency virus (HIV) infection, multidrug-resistant tuberculosis and viral hemorrhagic fevers among the workers during the last few decades [4, 5]. Published reports have given data of laboratory-associated cases of infection like as brucellosis, Q-fever, typhoid, AIDS, tuberculosis, hepatitis and so on. There are various occupations which lead to exposure to harmful microbiological wastes like pharmaceutical industries, diagnostic laboratories, microbiological research laboratories and also funeral homes that generate a huge amount of waste. The assessment of public and

Nazia Tarannum, Department of Chemistry, Chaudhary Charan Singh University, Meerut 250004, India, e-mail: naz1012@gmail.com
Ranjit Hawaldar, Centre for Materials for Electronics Technology, Pune, India

https://doi.org/10.1515/9783110517736-008

occupational health risks related with these wastes has been always undervalued. Therefore, while dumping microbiological wastes, working personnel should take special care of the content and survival of microorganisms in such wastes, which include pathogens in municipalities and healthcare waste. There are few most common components of biological wastes as follows:

- Cultures of infectious agents, including antigens, vaccines production processes and sera developed waste in research or clinical lab
- Pathological wastes removed during surgery or autopsy like organs, tissues and body fluids and parts
- Blood and components of blood (e.g., serum and plasma)
- Surgical items or disposed cloth contaminated with blood
- Waste from animals like body fluids, carcasses, animal body parts from research institutes or veterinary clinics
- Zoonotic wastes like infected animal urine, feces or other bodily fluids

The chapter focuses on work-related infectious diseases, which include diseases from broad-spectrum pathogenic agents like bacteria, viruses, fungi or parasites through human, animal and strong contact across different occupational groups. It is an estimation that every year the employees die from microbial infections obtained occupationally, but the primary challenge faced by a high rate of mortality is that the disease has not been properly assayed. Lack of awareness among the workers regarding pathogen handling increased the risk. Further, the policies made to control hazards are not adequately applied. The chapter discusses the role of microbiology in different occupational settings, legislation and safety policies in action for cause and biosafety measures to adapted and practiced in microbiology labs.

8.2 Role of microbiology hazards in various occupations

The hazardous components of microbiological waste cause microbiological, chemical and physical risks to public and people involved in these hazardous components treatment, disposal and handling. Unlike, its beneficial effects, microbiology is also proving to be a threat to the health of working professionals exhibited to its effect by any means.

The following professionals in the different setting are most likely to encounter microbiological waste during its handling, generation, disposal and treatment:

- Healthcare personnel: pathologists and doctors
- Laboratory workers: scientist, researchers, forensic toxicologists, lab technicians
- Waste workers
- Substance abusers
- General public

In laboratory-acquired infection, professionals working in the lab, especially those in microbiology are prone to infections. The risk assessment depends on some factors like the host immune system, toxic exposure dose, agent virulence, mechanism of exposure and use of personal protective equipment. The generation of data of risk is challenging, as there is no efficient system to monitor the reports related to the number of laboratory-related infections or exposures. The Disease Prevention and Control Centers have convened a system committee to highlight these issues to provide guidelines for exposure risk. To discover vaccines for deadly diseases in the laboratory, scientists are creating conditions in labs, which are incurable. The reports from Hong Kong in 1997 have given an account of the death of two dozen people due to theH5N1 avian flu. There has been a case of infection of severe acute respiratory syndrome (SARS) in 2003 of a Singaporean lab worker during his research work. In 2004, a Russian scientist in Siberia fortuitously stuck herself with Ebola-contaminated needle and died. There have been cases of misuse of SARS virus like loss of vials containing SARS virus, which cause bleeding under the skin, body orifices or internal organs. People involved in medical field are making an effort to eliminate these diseases, and to do so they need to preserve the sample for prospective studies. This preserving of incurable disease virus drove them to direct or indirect exposure to the hazard. In West Bengal, India, poor management of healthcare waste caused medical specialists to reuse glass syringes instead of mandatory single-use plastic syringes. Reports suggest that reuse of unsterilized needle causes HIV infections hepatitis B and hepatitis C [6].

The researchers in the field of biochemistry, biotechnology, microbiology, zoology, bioinformatics or pharmaceutical sciences and so on are under constant exposure to various microorganisms. In the course of their study processes, they handle and work with different agents. Health risks get high during waste treatment through the release of toxic pollutants and direct exposure of the workers involved in it. In waste disposal locations, employees handled untreated waste manually and were exposed to blood or body fluids wastes or airborne pathogens. These waste components may act as a source to create infectious aerosols containing etiologic agents of human or animal diseases. The biological wastes from laboratories may include cultures of infectious agents [7–9]. Another source of infectious aerosols is animal and human tissues, animal bedding materials, body parts, which act as a potential microbiological hazard [10, 11].

Forensic toxicologists and laboratory technicians involved in cutting viscera of case samples for scientific investigation are at constant risk of exposure to biological wastes, untreated spleen, intestine, liver, blood and so on. Constant exposure to foul smell or physical contact with infected or discarded wastes leads to dangerous health risks.

Some reports from the USA have shown tuberculosis outbreak among employees at an organic waste treatment area [12]. The poor waste management practices, limited availability of immunization and lack of personal protection have posed a significant risk to the low-income countries. A survey data in 1990 demonstrated that Italy had higher cases of hepatitis in municipal workers than those found in general public [13]. Collins and Kennedy in their study of healthcare waste reported that 2% of residues of blood were positive for viruses of hepatitis [8]. Highest incidents of needle stick injuries have been reported in the United Kingdom by a survey among nurses, physicians and housekeeping staff [14]. Researchers working on some viruses are contaminated by the needles. Sometimes, in culture labs, cells stored in nitrogen tanks may get leaked in the cylinder due to some reasons. Due to this, often the researchers are exposed to the contaminated nitrogen vapors.

The first outbreak of SARS occurred in Vietnam in 2003, where most of the patients affected by the disease were hospital staff. Urbani, an epidemiological and clinical expert from the World Health Organization (WHO), was called off to study the severity of the disease [15]. The close contact with the lethal virus was concluded as the cause of sudden death of Urbani. In West Africa, there was a significant outbreak of Lassa fever, a rodent-borne disease transmitted by direct contact with infected blood, contaminated needles and syringes or abnormal sanitary conditions [16].

8.3 Occupational zoonotic diseases

A primary cause of zoonotic diseases is a pathogen transmitted from water, soil or animal source which is contaminated with infected urine of animal, other bodily fluids or feces. Employees involved in handling of animal like technicians, farmers, abattoir workers, veterinarians and so on are prone to zoonotic pathogens [17]. Changing farming conditions and socioeconomic conditions have led to a phenomenal growth of the disease. A well-known lethal disease is Anthrax caused by *Bacillus anthraces* spores. The infections of human are cutaneous due to highly resistant spores, which enter via skin wounds [2]. Anthrax has been a threat due to its use as a biological weapon [18].

8.4 Legislation and safety policies

Many countries have set safety standards for licensing laboratories for clinical purpose to work with infectious agents, packaging, disposal of such materials and exposure to workers. For this WHO, the National Institute for Occupational Safety and Health, Occupational Safety and Health Administration (OSHA), the Centers for Disease Control, the Advisory Committee on Dangerous Pathogens, come into action. The personnel protection from risks of exposure to biologicals at work amended by the Directives 93/88/EEC, 95/30/EC, 97/59/EC, 97/65/EC is under European Parliament and Council of European Union. The council directive has set safety policies for laboratories, which involve biological agents for diagnostic purposes, research and development to minimize the risk of infection. The Directive actively assesses risk measures by an infectious agent, their quantities and susceptibility of the infectious agent to laboratory workers. Using the above directives viruses, bacteria and fungi are classified into active risk groups. According to Directive 2000/54/EC, measures of containment for CL-2, CL-3 and CL-4 laboratories have been classified based on risk groups. The directive also proposed a particular use of microbiological safety cabinets in the laboratory to reduce exposure of aerosols while handling pathogenic agents [19–21].

In the USA, the protection policy of the OSHA regulations, the Department of Labor, patient confidentiality laws and Americans with Disabilities Act of 1990, offer health services from the medical support team to workers prone to exposure to microbiological agents [8, 22–24]. High-income countries have set a few safety policies for disposal of these biowastes. According to proposed policy, treatment and disposal of waste liquid blood are done through the sewage system. By diluting and treatment procedure, concentration and viability of blood-borne pathogens are reduced [24]. In Europe, largest microbiology society is Society for General Microbiology, with a membership of 5,300 worldwide from universities, industry, research institutions, health and veterinary services.

8.5 Biosafety measures in microbiological laboratories

The goal of microbiological laboratories in medical and research settings should focus on a safe and healthy workplace. Reducing possibilities for exposure, detecting and treating exposures at an early stage and using proper safety precautions are important. Issues of occupational health and security in the area are a shared responsibility of workplace personnel, principal investigators, safety specialists and employers. Awareness of the potential worksite health hazard and a proper medical support service should be designed to neutralize the harmful effect.

A few essential medical support services encouraged are outlined below [25].

Measures to control the possible route of entry of pathogens into the body are discussed below:

1. Oral route: Use of proper apron/gloves, eye protection and particle mask should be part of wear. Eating/drinking/smoking/gum chewing should be prohibited. Regular handwashing with soap/water should be encouraged.
2. Respiratory Route: Filter mask (HEPA) should be used, working areas should through a regular disinfectant check.
3. Skin membrane: Skin cuts cause the transmission of rabies, rabbit fever, tetanus and so on. Protective clothing like gloves, boots and coveralls should be encouraged.

The following biosafety measures to be followed by personnel at the risk of exposure:
- Preplacement medical evaluations
- Vaccines
- Medical evaluations periodically
- Proper disposal of healthcare wastes

8.5.1 Preplacement medical evaluations

Personnel who are at constant exposure to pathogens in a pathology or laboratory should get a medical evaluation at a constant interval. Healthcare experts should inspect the workplace. A medical history, allergies, current medications of workers must be evaluated as required. Employees should be familiar with possible health hazards in a work area and notified of measures. To avoid the accidental exposure, employers should collect and store a specimen serum before the initiation of work. Non-immune pregnant female or immune-deficient employees may face devastating consequences from exposure to cytomegalovirus or toxoplasmosis; therefore, special care should be monitored in their case [26].

8.5.2 Vaccines

Periodically, workers at risk of occupational exposure should be provided with vaccines for protection against infectious agents [27–31].

The investigational vaccines for Venezuelan equine encephalitis virus, eastern equine encephalomyelitis virus, Rift Valley fever viruses and Western equine encephalomyelitis virus are administered on-site to the workers who are exposed to the particular pathogens, occupationally.

8.5.3 Medical evaluations periodically

Periodic laboratory testing should facilitate professionals for detecting pre-clinical evidence of an occupationally acquired infection. The modes of medical presentation and transmission of diseases through occupational exposure may differ from infections, which are acquired otherwise. Strategies should be taken care of in advance for responding to Biohazard exposures. A printed summary of laboratory SOPs should include guidelines for immediate medical response to specific exposures.

The format of medical description of the injury should involve:
- The infectious agent
- The exposure route (skin, aerosol, mucous membranes, percutaneous etc.)

- Incident time and site
- Personal protective tools used at an event of injury
- Details of first aid provided (e.g., lapse of time from exposure to treatment)
- Patient medical history records

In the case of any accidental risk, each incident should be considered for cases and review of current strategies. There are certain infections, which may prove lethal to workers and may produce immediate health concern for those in the vicinity by transmission like an infection by BSL-4 agents (bio safety level).

8.5.4 Proper disposal of healthcare wastes

The occupational health risks from microbiological agents found in healthcare wastes may be reduced by taking the following measures:
- Sharps like needles and syringes, Pasteur pipettes, scalpel blades and so on may cause injury and act as an agent to spread of infections. Healthcare waste should be poised and stored in puncture-proof containers and should not be reused.
- Blood and products of blood like discarded plasma and serum are reservoirs of infectious agents. These wastes should not merge with municipal wastes and appropriately treated before disposal.
- Infectious agents from cultures and stocks of research and clinical laboratory must be disposed of properly.

8.6 Challenges faced by occupational hazard

The original problem faced by work-related infections include incorrect diagnosis and under-reporting, which may be the result of inadequate access to professional healthcare. Health experts in the workplace may not have access to proper diagnostic tools, or workplace may not be monitored properly [3]. Further, the risk at the site is a resurgence of anonymous microorganisms and new disease [32]. Another growing challenge is blended exposures that occur in the workplace, which may lead to problems in identifying the causative agent [33].

8.7 Conclusion

Nowadays, expanding awareness has led laboratories to take critical biosafety standards for facilities, practices and equipment to prevent possible hazards to staff and community. In recent years, some countries have shown concern for biosafety measures as the upcoming harmful microorganisms can act as a potential bioweapon. Literature reviews on healthcare wastes associated microbiological hazards have shown many areas of concern leading to pathogens transmission to waste handling workers and the public. Medical wastes should be classified according to their source of origin, risk factors and typology. It is crucial to take some necessary steps to reduce healthcare wastes and recycle them into environment-friendly wastes.

References

[1]   Chung YK, Ahn Y-S, Jeong JS, Occupational infection in Korea, J Korean Med Sci, 2010, 25 (Suppl), S53–S61.
[2]   Lim VKE, Occupational Infections, Malaysian J Path, 2009, 31(1), 1–9.
[3]   Haagsma JA, Tariq L, Heederik DJ, Havelaar AH, Infectious disease risks associated with occupational exposure: A systematic review of the literature, Occup & Environ Med, 2012, 69, 140–46.
[4]   Sepkowitz KA, Eisenberg L, Occupational deaths among health care workers, Emerg Infect Dis, 2005, 11(7), 1003–08.
[5]   Sagoe-Moses C, Pearson RD, Perry J, Jagger J, Risks to health care workers in developing countries, N Engl J Med, 2001, 345, 538–41.
[6]   Kane A, Lloyd J, Zaffran M, Simonsen L, Kane M, Transmission of hepatitis B, hepatitis C, and human immunodeficiency viruses through unsafe injections in the developing world: Model-based regional estimates, Bull World Health Org, 1999, 41, 151–54.
[7]   Reuters Online, 19 Jun, 2000, Smallpox, Russian Children Affected By Dumped Vaccines.
[8]   Collins CH, Kennedy DA, The microbiological hazards of municipal and clinical wastes, J Appl Bacteriology, 1992, 73, 1–6.
[9]   Centers for Disease Control and Prevention. Guidelines for preventing the transmission of *Mycobacterium tuberculosis* in health-care facilities, Morb Mort Weekly Rpt, 1994, 43, RR–13.
[10] Gershon RR, Vlahov D, Escamilla-Cejudo JA, *et.al*, Tuberculosis risk in funeral home employees, JOccup Environ Med, 1998, 40, 497–503.
[11] Sterling TR, Pope DA, Bishai WR, Harrington S, Gershon RR, Chaisson RE, Transmission of *Mycobacterium tuberculosis* from a cadaver to an embalmer, New Eng J Med, 2000, 342, 246–48.
[12] Weber AM, Boudereau Y, Mortimer VD, Health Hazard Evaluation Report 98-0027-2709, Stericycle, Inc, Morton, Washington, Cincinnati, National Institute for Occupational Safety and Health, 1999.
[13] Kanitz S, Franco Y, Roveta M, Patroone V, Raffo E. Sanitary landfilling: Occupational and health hazards. Proceedings of Sardinia 91, Third International Landfill Symposium, October 1991.
[14] Morgan DR, Needlestick and infection control: Policies and education, AIDS Letter, 1999, 71, 1–4.
[15] Chee YC, Heroes and heroines of the war of SARS, Singapore Med J, 2003, 44, 2221–44.
[16] Fisher-Hoch SP, McCormick JB, Towards a human Lassa fever vaccine, Rev Med Virol, 2001, 11, 331–41.
[17] National Institute for Communicable Diseases (NICD): Annual Report 2011-2012. Johannesburg: NICD; 2012.

[18] Dutkiewicz J, Cisak E, Sroka J, Wójcik-Fatla A, Zaj Ä, Biological agents as occupational hazards–selected issues, Annals Agri & Environ Med, 2011, 18(2), 286–93.

[19] COUNCIL DIRECTIVES 90/679/EEC (OJ L 374, 31. 12.1990,p.1), 93/88/EEC (OJ L 268, 29. 10.1993,p.71), 95/30/EC (OJ L 155, 6. 7.1995,p.41), 97/59/EC (OJ L 282, 15. 10.1997,p.33), 97/65/EC (OJ L 335, 6. 12.1997,p.17) and 2000/54/EC (OJ L 262, 17.10.2000, p.21)

[20] Categorization of Biological Agents according to Hazard and Categories of Containment, 4th, Advisory Committee on Dangerous Pathogens. HSEBooks, 1995 (supplement, 1998). *The Management, Design and Operation of Microbiological Containment Laboratories*.Advisory Committee on Dangerous Pathogens, HSE Books, 2001.

[21] Laboratory Biosafety Manual, 2nd, World Health Organization, 1993, revised 2003.

[22] Owen K, Leese L, Hodson R Control of Aerosol (Biological and Non-Biological) and Chemical Exposures and Safety Hazards In Medical Waste Treatment Facilities. Final Report 1997.

[23] Centers for Disease Control and Prevention, NIOSH Cincinatti. OH, Rutala WA, Mayhall CG, Medical waste, Inf Cont Hosp Epi, 1992, 13, 38–48, Pruss A, Giroult E, Rushbrook P (eds). Safe management of wastes from health-care activities. Geneva, World Health Organization, 1999.

[24] Collins CH, Kennedy DA, The Treatment and Disposal of Clinical Waste, Leeds, H&H Scientific, 1993.

[25] Menckel E, Westerholm P editors, Evaluation in Occupational Health Practice, Oxford, Butterworth-Heinemann, 1999, Levy B, Wegman DH, editors. Occupational health: recognizing and preventing work-related disease and injury, 4th ed. Philadelphia: Lippincott Williams & Wilkins; 2000.

[26] National Research Council (US), Committee on Occupational Health and Safety in the Care and Use of Nonhuman Primates, Institute for Laboratory Animal Research, Division on Earth and Life Studies, Occupational Health and Safety in the Care and Use of Nonhuman Primates, Washington, DC, National Academy Press, 2003.

[27] Atkinson WL, Pickering LK, Schwartz B, *et al.*, General recommendations on immunization. Recommendations of the Advisory Committee on Immunization Practices (ACIP) and the American Academy of Family Physicians (AAFP), MMWR Recomm Rep, 2002, 51, 1–35.

[28] Centers for Disease Control and Prevention. Update on adult immunization. Recommendations of the immunization practices advisory committee (ACIP), MMWR Recomm Rep, 1991, 40, 1–94.

[29] Centers for Disease Control and Prevention. Immunization of health-care workers: Recommendations of the Advisory Committee on Immunization Practices (ACIP) and the Hospital Infection Control Practices Advisory Committee (HICPAC), MMWR Morb Mortal Wkly Rep, 1997, 46, 1–42.

[30] Centers for Disease Control and Prevention. Use of vaccines and immune globulins in persons with altered immunocompetence. Advisory Committee on Immunization Practices (ACIP), MMWR Morb Mortal Wkly Rep, 1993, Apr, 9 42(RR-04), 1–18.

[31] Centers for Disease Control and Prevention. Update: Vaccine side effects, adverse reactions, contraindications, and precautions. Recommendations of the Advisory Committee on Immunization Practices (ACIP), MMWR Recomm Rep, 1996, Sep, 6, 45(RR-12), 1–35, Erratum in: MMWR Morb Mortal Wkly.

[32] Snashall D, Patel D, ABC of Occupational and Environmental Medicine, 2012, Chichester, UK, John Wiley & Sons Ltd.

[33] Eduard W, Heederik D, Methods for quantitative assessment of airborne levels of non-infectious microorganisms in highly contaminated work environments, Am Ind Hyg Assoc J, 1998, 59, 113–27.

Preeti Kumari Sharma, Paavan Singhal
Chapter 9
Medical waste management

Abstract: The waste generated from medical labs, hospitals, various centers and re-search institutes constitute harsh and potent health hazards. It is necessary to regulate the waste generation, management of generated waste and possibility of recycling or reuse of generated waste. The medical waste management systems should be strictly controlled. Biological hazards and risks are usually from human samples, tissues and inanimate objects. To assess risk accurately, biosafety programs are must. This chapter deals with the medical waste management including types of medical waste, risks associated, waste treatment and management strategies.

Keywords: medical waste, health hazard, containment level, biosafety, chemical agents, medical waste management

9.1 Introduction

The waste from medical centers and hospitals constitute particularly harsh and potent health hazards, due to this reason, it is necessary that waste generated from medical center is managed separately from general solid waste. The waste management systems are more strictly controlled. Biological hazards and risks are usually from human samples, tissues and inanimate objects. To assess risk accurately, biosafety programs are must. It is thus essential that safety measures are enforced at all times so as to reduce the chance of exposure of pathogens and possibility of infections to the staff of the laboratory. Biosafety will also facilitate to decrease accidental discharge of such pathogens into the immediate setting.

In healthcare settings and hospitals, antiseptics and disinfectants are widely used to control infections. Antiseptics and disinfectants contain variety of active chemical agents which have the ability to control and inhibit the growth of potent microbes. Commonly used antiseptics and disinfectants are alcohols, phenols, iodine, chlorine and so on. The constituents of antiseptics and disinfectants exert their antimicrobial effect by denaturing the microbial protein, disrupting of cell membrane and inhibiting the action of enzymes.

Preeti Kumari Sharma, Department of Microbiology, Government Medical College and Hospital, Jammu. e-mail: priitisharma.micro@gmail.com; Department of Biotechnology, Maharishi Markandeshwar (Deemed to be University), Mullana-Ambala
Paavan Singhal, Department of Biotechnology, Maharishi Markandeshwar (Deemed to be University), Mullana-Ambala

https://doi.org/10.1515/9783110517736-009

Biomedical or hospital waste is far more dangerous than domestic waste because it contains infectious and hazardous materials that can further infect or harm patients and hospital staff in several ways. If biomedical waste is kept untreated, it would start fermenting and also attract flies which make the place dirty and unhygienic. Sharp objects like syringes, needles or broken glasswares can easily cause injury and infection. So, management of biomedical waste is very important. Usually waste tends to accumulate in filthy and dirty surrounding. The preliminary requirement for effective waste management is neat and clean environment. The hospital premises should be kept in clean and hygienic conditions to avoid nosocomial infections. The waste management includes various different steps: reduction, reuse, segregation, storage, transportation and treatment. Many methods of waste treatment are available on the basis of items disposed in the waste. If possible, final disposal place may be away from crowded areas to avoid the risk of further infections. Some of the methods are chemical disinfection, incineration, autoclaving and so on.

9.2 Definitions

Hospital waste: All waste, biological or non-biological that is discarded and not intended for further use is called as hospital waste.

Biomedical waste: The waste that is generated in the process of immunization, diagnosis or treatment of patients in hospitals is called as bio-medical waste.

9.3 Type of medical waste

Hospital waste includes all waste (biological and nonbiological) which are discarded and disposed. Categories of waste from medical centers are shown in Figure 9.1.

The waste generated in hospitals can be classified in two main categories:
– **Clinical (medical) waste**
– **Nonclinical (general) waste**

1. **Clinical (medical) waste:** Clinical waste consists of 10–25% of the total waste generated in the hospital. It is the infectious waste which presents serious risk to human health and can cause various infectious diseases. Clinical waste can be classified into the following classes:
   – Infectious waste: It includes waste from laboratory culture, tissues, used dressings and so on.
   – Pathological waste: It includes waste from blood, blood products, human fetus, body fluids, body parts and so on.

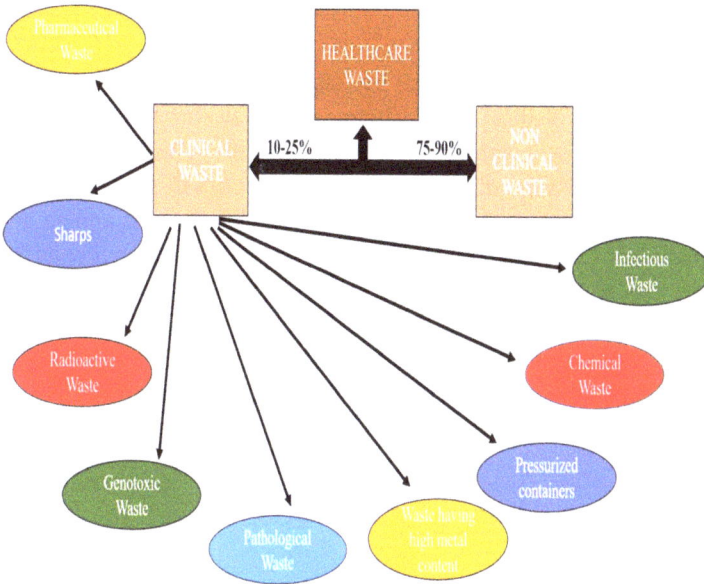

Figure 9.1: Categories of waste from medical centers.

- Chemical waste: It includes waste from unused and surplus degraded reagents, chemicals from diagnostics, research work and so on.
- Pharmaceutical waste: It includes waste from unwanted and expired drugs and so on.
- Sharps: It includes waste from sharp objects like syringes, needles, blades and so on.
- Radioactive waste: It includes waste from the actual radiation source in radiotherapy unit, medical instruments contaminated with certain isotopes or certain type of laboratory work and so on.
- High heavy metal contents: It includes waste from the broken thermometers, blood pressure gauges and so on.

Usually all types of medical waste are harmful and infectious.
2. **Nonclinical (general) waste:** General or nonclinical waste is noninfectious waste generated in the hospital premises. It comprises between 75% and 90% of the total amount of waste that is generated in the hospitals. For example, items like nappies, incontinence pads and so on. This type of waste does not cause any harm [1].

9.4 Accompanying risks

General, nonclinical, noninfectious waste poses less significant hazards as compared to medical or infectious waste. Medical or infectious waste may cause serious infectious diseases and risks. The most common associated risk is the spread or transmission of infectious diseases like viral diseases (HIV and hepatitis B) through direct contact with the infected wastes like clinical specimen, syringes, needles and so on. To avoid risk, all specimens specifically blood must be disposed in leakproof bags for transportation. Proper use of masks, gloves while handling specimen is must. Decontaminate the laboratory after completion of diagnostic and research procedures. Use of biosafety cabinets (BSCs) is a must in diagnostic procedures [2].

Nondirect risk includes infections and diseases spread by vectors, water pollution and pollution of environment.

9.4.1 Channeling pathways

The initial phase of waste management is to recognize the disease transmission pathways. Common potential pathways between medical waste and public areas are described as under:

– Direct contact
– Indirect contact via vectors
– Transmission by air
– Transmission by water
– Pollution of the environment

Risks, pathways and hazards of medical waste are described in Table 9.1.

Table 9.1: Risks, pathways and hazards of medical waste.

Risks	Pathway	Hazards
Infection/disease contraction	Direct or indirect contact via carrier	Infectious waste may transmit infection or disease through direct and indirect contact.
Cuts	Direct contact	Encounter with sharp waste like syringes, needles may cause cut and provide entry of pathogen in the body.
Infective medical care	Direct	Accidental consumption of expired drugs
Skin irritation/ burns	Direct or indirect contact via carrier	Radioactive waste and toxic chemicals

Table 9.1 (continued)

Risks	Pathway	Hazards
Cancer	Direct or indirect contact via carrier	Radioactive waste
Pollution of air, water and environment	Direct or indirect contact with polluted air, water or atmosphere	Toxic chemicals waste, pharmaceutical waste and heavy metal contact

To minimize the risk of medical waste, it is very important to break potential pathways associated with the transmission of medical waste.

The particular groups of population at risk from medical waste are:
- Clinicians
- Laboratory staff
- Medical waste workers
- Waste picker laborers
- Patients
- Children (playing near disposal sites)
- Drug addicts (using disposal syringes)

Proper education and awareness should be given to workers who come directly in contact with medical waste by organizing workshops and conferences.

9.5 Minimizing risks of medical wastes

High hazardous risk from medical waste can be minimized by
- Proper labeling of hazardous wastes
- Use of incineration to reduce hazards.
- To separate different medical wastes at the time of their generation.
- Disinfecting the medical waste before disposing.
- Disposing different types of medical waste via proper disposal systems to avoid mixing of medical waste.

Direct contact between personnel and medical waste can be prevented:
- By education and awareness of hazards of medical wastes
- By restricting access to medical waste pits
- By providing personal protective clothing like gowns, masks, gloves and so on
- By implementing healthy practices while dealing with waste
- Protecting water source from contamination to avoid infections

- Providing better facilities to the workers those who deal directly with waste
- Immunization of health care workers dealing with medical waste [3]

9.5.1 Infectious microorganisms associated with hospital medical waste

The most common microorganisms which cause infections or diseases from medical waste are as follows:
- HIV
- Hepatitis B
- *Brucella*
- *Mycobacterium tuberculosis*
- *Bacillus anthracis*
- *Salmonella*
- *Shigella*
- *Coccidioides immitis*

9.6 Hazard group and containment level

9.6.1 Hazard groups

Biological substances that cause threat to the living organisms, particularly human beings are called as biological hazards or biohazards. It is comprised of medical waste, microbes or toxin which can affect the health of humans. In 1966, Charles Baldwin developed a symbol for biohazards and this symbol is commonly used as a warning [4].

9.6.2 Levels of biohazard

Various microbial infections and diseases are classified according to the levels of biohazards set by the Centers for Disease Control and Prevention (CDC). The levels of biohazards are arranged from Level 1 to Level 4. Level 1 means minimum level of risk to human health whereas Level 4 is being at high risk to human population. Facilities in laboratories of hospitals and research institutes are categorized as biosafety level (BSL) from BSL-1 to BSL-4. BSL-1 laboratories deals with microbes which pose little or almost no risk to human health whereas BSL-4 deals with dangerous microbes which cause fatal infections to humans [5].

– **Biohazard level 1:**

At biohazard level 1, medical wastes having biohazardous materials which are less toxic and cause minimum risk to human lives are included. Biohazard level 1, commonly deals with bacteria like *E. coli, B. subtilis* and viruses like hepatitis, chicken pox virus and so on.

– **Biohazard level 2:**

At biohazard level 2, biohazardous materials are little more toxic than biohazard level 1. At this level, common dealing is with bacteria like *Salmonella* and viruses such as hepatitis A, B and C; some influenza A strains, dengue virus, HIV and so on. Daily diagnostic laboratory work on these bacteria and virus can be done safely by using BSL-2.

– **Biohazard level 3:**

At biohazard level 3, biohazardous material contains microbes which have the ability to cause serious diseases in humans, although vaccine and treatment is easily available for these microbes. Bacteria like *M. tuberculosis, B. anthracis* and viruses like West Nile virus, SARS virus, Hantaviruses, yellow fever and so on are dealt at BSL-3.

– **Biohazard level 4:**

At biohazard level 4, there is dealing with highly potent and hazardous microbes which has the ability to cause fatal disease in human beings. There is no availability of vaccines and treatment against these highly pathogenic microbes. Pathogens like Marburg virus, Ebola virus, Crimean–Congo hemorrhagic fever and Lassa fever virus are dealt at biohazard level 4.

While working or dealing with highly pathogenic microbes, one should take proper precautions like a proper UV-room, a vacuum room, multiple showers at the time of entry and exit and wearing a positive pressure personnel suit [6].

9.6.3 Laboratory biosafety level criteria

Biological safety levels are a series of protection levels and the levels are arranged in ascending order according to the degree of safety provided to the laboratory personnel, as well surrounding environment and community. These levels are classified from one to four, according to the microbes that are being worked on in any laboratory. For exhibiting particular controls for the containment of microbes or biological agents, CDC sets biosafety laboratory levels from BSL-1 to BSL-4. BSLs are shown in Figure 9.2.

BSLs are very essential because it allows safe practice with pathogenic microbes by providing all type of specialized safety equipment within laboratory premises [7].

The following is an explanation of each BSL in detail:

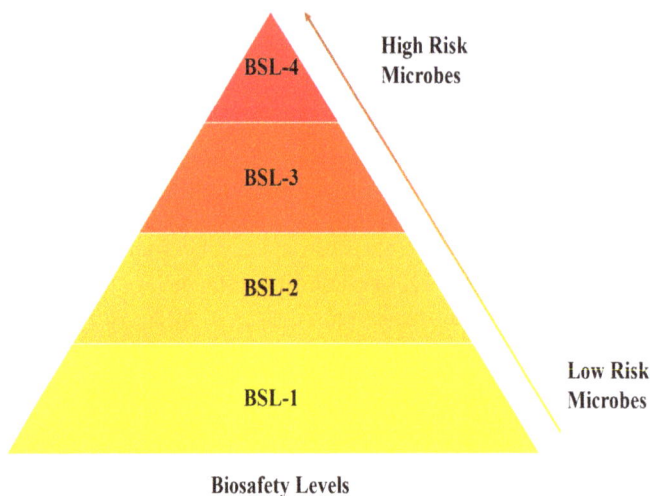

Figure 9.2: Biosafety levels (BSL).

9.6.4 Biosafety level 1

Out of the four BSLs, BSL-1 is the lowest that deals with low or minimum risk microorganisms that pose no threat to human population. BSL-1 is suitable for work with nonpathogenic strains and cause minimum hazard to laboratory personnel, environment or community. It is not essential to isolate laboratories from surrounding facilities at BSL-1. The following facilities are applied to BSL-1.

1. Laboratory workers should wash their hands after dealing with hazardous materials.
2. Smoking, drinking, eating and storing food is not allowed in the laboratories.
3. Only mechanical pipetting is allowed.
4. Proper policies must be implemented for the safe handling of glasswares, scalpels, needles and pipettes.
5. All the procedures should be performed carefully so as to minimize the splashes.
6. Laboratories should be decontaminated by using appropriate disinfectant after the completion of work.
7. Laboratory personnel should wear protective equipment, such as gloves, masks and a lab coat or gown.
8. Access to the laboratory should be minimized wherever and whenever infectious agents are present.
9. At the entrance of the laboratory, signs of biohazard symbol must be posted. The name and telephone number of the laboratory supervisor should also be posted at the entrance to avoid any emergency.

10. A proper and effective integrated pest management program should be conducted at regular intervals for the better functioning of laboratories [8].

9.6.5 Biosafety level 2

BSL-2 is suitable for work with pathogenic agents that constitute mild hazards to the lab worker and the environment. Examples of microbial agents worked with in BSL-2 laboratories include HIV as well as *Staphylococcus aureus*. It is different from BSL-1 in that:
1) Laboratory workers should work under the supervision of scientists competent in handling the procedures related with pathogenic microbes.
2) Access to the laboratory should be restricted only to laboratory personnel, when work on pathogenic microbes is being conducted. Access to a BSL-2 laboratory is far more restrictive than to BSL-1 laboratory.

The following facilities are applied to BSL-2 laboratories:
1. Proper medical surveillance and immunizations should be provided to laboratory personnel as they are dealing with pathogenic microbes.
2. Biosafety manual for laboratory must be prepared and followed properly.
3. During gathering, holding, processing, storing or transportation of microorganisms, they must be kept in a leakproof container to avoid any spill.
4. All equipment in the laboratory should be decontaminated on daily basis and after splashes or spills of infectious agents.
5. Proper use of personal protective equipment is a must in the laboratories which includes lab coats, gloves, eye protection and face shields.
6. All the accidents related to exposure of infectious material must be reported to laboratory supervisor and immediately evaluated according to the procedures set in laboratory biosafety manual.
7. The laboratory has self-closing and lockable doors.
8. An eyewash station or sink should be set in the laboratory.
9. Appropriate policies must be developed and implemented for the safe handling of sharps, such as scalpels, needles, pipettes and broken glassware.
10. Proper attention should be paid to control the routes of exposure, including advanced techniques for handling the contaminated sharp objects [9].

9.6.6 Biosafety level 3

BSL-3 builds upon BSL-2. Teaching, research, diagnostics, clinical laboratories are applicable to BSL-3. At BSL-3, laboratory procedures are performed with highly pathogenic bacteria and viruses which pose serious health hazardous diseases in humans through the inhalation. At BSL-3, work deals with viruses like yellow fever, West Nile virus and bacteria like *M. tuberculosis*.

These microbes are highly pathogenic and work related to them must be controlled and regulated by proper government agencies. Appropriate immunization and medical surveillance should be provided to laboratory personnel who are working with potent microbes to reduce the chances of infections.

The following facilities are applied to BSL-3:

1. All the protocols in the laboratories involving the manipulation or alteration of microbes must be conducted within a BSC usually preferred Class II or Class III.
2. In the laboratory, workers should wear protective laboratory clothing with a solid-front or wrap-around gowns, scrub suits and so on. Laboratory clothing which is reusable must be decontaminated before wash.
3. Face and eye protection (mask, goggles and face shield) should be used while working against splashes of infectious materials. All masks must be properly disposed after use. Gloves must be used to protect hands from exposure to hazardous materials and to minimize the risk of infections.
4. Laboratory doors must have self-closing locks. Laboratory access should be restricted. Access in the laboratory should only be through two self-closing doors.
5. A sink or handwashing station must be present in the laboratories. The taps in the sink must be automatic and should be placed near the exit door of laboratories.
6. All laboratory windows must be closed.
7. In laboratories, BSCs should be located away from the doors and installed to avoid fluctuations of the room air supply.
8. High-efficiency particulate air (HEPA) filters must be used to protect vacuum lines and these filters must be replaced whenever required.
9. BSL-3 laboratories require ducted airflow ventilation systems. It must provide unidirectional airflow by withdrawing air in the laboratory from clean areas toward certain contaminated areas. Laboratories should be designed in such a way that under failure conditions, the airflow never gets reversed. The system must provide sustained directional airflow.
10. All laboratory wastes should be decontaminated in the laboratory facility by using validated decontaminated methods [10].

9.6.7 Biosafety level 4

BSL-4 laboratories deal with highly dangerous and pathogenic microbial agents that cause life-threatening diseases to human population. BSL-4 laboratories consist of work with pathogenic and exotic microbes for which no vaccines or treatments are available, for example, Ebola viruses and Marburg viruses. BSL-4 laboratories are found to be very rare. BSL-4 laboratory are design in such a manner to prevent dissemination of microbes into the environment. All laboratory workers and supervisors must be competent in handling exotic and highly pathogenic microbes. Usually, there are two types of BSL-4 laboratories:

1. Cabinet laboratory: It involves manipulation and alteration of microbes that must be performed in a class III BSC.
2. Suit laboratory: It requires that laboratory workers must wear a positive pressure supplied air-protective suit to avoid infections.

The following facilities are applied to BSL-4.

1. All the BSL-4 laboratories must be set in highly isolated or demarcated zone within a building or either in a separate building. Air supply, alarms, life support, security systems and entry and exit controls, their monitoring and control should be on an uninterrupted power supply to avoid unnecessary mishaps.
2. All laboratory windows must be closed and must be heat- and break-resistant.
3. Appropriate ventilation system is needed in the laboratory.
4. Proper communication systems should be provided to communicate between the laboratory and the outside for emergency communication.
5. The decontamination process is very important, especially for liquid wastes. Liquid waste must be validated both physically and biologically. Biological validation must be performed annually or as prescribed by institutional policy.
6. Laboratory personnel must wear appropriate personal protective equipment while working in the laboratory which include full body, air-supplied, positive pressure suit. All laboratory personnel are directed to change clothing before entering and shower upon exiting the laboratory to avoid unnecessarily contamination.
7. Decontamination of all materials of laboratory before leaving.
8. Strict control access is given only to those personnel who are permitted to be in the BSL-4 laboratory.
9. All HEPA filters must be tested, validated and certified annually. All air from exhaust, fumigation and decontamination shower should pass through two HEPA filters arranged in series, before outlet.
10. Infectious material spills should be decontaminated and cleaned by competent professional staff. A spill procedure should be displayed within the laboratory so that it helps during the time of emergency [11].

9.7 Antiseptics and disinfectants

9.7.1 Antiseptics

The antimicrobial substances that are applied on the living surface to stop or slow down the growth of microbes are called as antiseptics. These antiseptics inhibit the microbial infections. These antiseptics are very effective against microorganisms like bacteria, fungi and others. Antiseptics and antibiotics are very much different because the antiseptics are mainly used as topical agents whereas antibiotics work directly inside the living body. In the nineteenth century, the research and development of antiseptics began [12].

9.7.2 Disinfectants

The substances which inhibit the microbial growth on inanimate objects are called disinfectants. These substances are not recommended for use on any living surface. The substances which have the ability to demolish vegetative forms of microbes which are pathogenic are known as disinfectants. The disinfectant has the capacity to reduce the microbes present on nonliving objects to minimal level. Characteristics of main disinfectants are discussed below in Table 9.2. [13].

9.7.3 Properties/characteristics of antiseptics and disinfectants

1. They have anti-microbial activity.
2. They are active against all types of the microbes.
3. These substances have quick action.
4. They are active against all skin exudates.
5. These substances are chemically stable.
6. They have the ability to destroy spores.
7. These substances produce minimum toxicity if absorbed.
8. They are nonsensitizing.
9. They do not cause any local irritation.
10. These substances are safe, easily available and cheap.
11. They are easy to apply.

9.7.4 Some of the commonly used antiseptics and disinfectants

– Formaldehydes
Formaldehyde is commonly used in laboratories and is effective against bacteria, spores and viruses. The solution of 10% aqueous solution of formalin is routinely used to sterilize instruments. The gas which is mainly used for fumigating, that is, sterilization of wards, sick rooms and laboratories is formaldehyde. So, clothing, bedding, furniture and books can be satisfactorily disinfected under properly controlled conditions.

When the gas is inhaled, it becomes irritant and toxic. The surfaces which have been disinfected by this agent may produce irritant vapors for some time after performing disinfection.

9.7.5 Glutaraldehyde

It has an action which is similar to formaldehyde. The toxic effects of glutaraldehyde is lesser when compared with formaldehyde. It is commonly active against bacteria (*M. tuberculosis*), viruses (hepatitis B), fungi and also spores. Glutaraldehyde is also used to disinfect delicate instruments having lenses. This antiseptic has broad spectrum and is also rapid in action [14].

– Alcohols
The most widely used disinfectant in the laboratories is alcohol. Ethyl alcohol and isopropyl alcohol are the most commonly used alcohols .These alcohols have the property to denature bacterial protein. They are commonly used as skin antiseptics in many setups. These alcohols generally show no action on spores and viruses. These are mainly used at 70–90% concentration in water to become more effective. Alcohols are also used to disinfect clinical thermometers. The alcohol which is commonly used as skin disinfectant is spirit or 70% ethyl alcohol. Alcohols are most commonly used for disinfection of surfaces. The only alcohol which is both toxic and inflammable is methyl alcohol vapors [15].

– Phenols
Phenols are one of the important and widely used disinfectants. The mode of action of phenol is by disrupting the microbial membranes which further results in inactivation of enzymes and precipitation of microbial proteins. Common examples of phenols are chloroxylenol, chlorhexidine and hexachlorophene. To prevent the infection of surgical wounds, Joseph Lister used phenols for the first time. They have microbicidal properties. They are effective against bacteria, *Mycobacterium* and fungi. They are inactive against viruses and spores. Corrosive phenols are mainly used for the disinfection of ward floors, bedpans and in discarding jars in laboratories.

Phenols like chlorohexidine are commonly used for skin disinfection or used as hand-wash. The 20% chlorhexidine gluconate solution is the most common and widely used disinfectant for general skin infections. The mixture of quaternary ammonium compounds such as cetrimide and chlorhexidine gluconate is recommended to get broader and stronger antibacterial effects. Chloroxylenols are very effective against bacteria, especially gram positive, and can be used as topical agents. It is less irritant than other phenols. To some extent, it is also effective against many gram-negative bacteria like *Pseudomonas* and *E. coli*. Chlorhexidine has good and effective antimicrobial activities against fungi and viruses. But, it is more irritant, toxic and corrosive to skin [16].

– Halogens

The Greek word "halos" means salt. So, halogen means salt former. Chlorine, iodine and bromine are three important halogens and act as microbicidal agents. They mainly act by forming protein halogen compound in living cells which cause damage of microbial enzymes. They are active against microbes, especially bacteria and spores. Chlorine reacts with water to form hypochlorous acid which is microbicidal. The chlorine compounds are hypochlorite, bleach and so on and iodine compounds are iodophores, tincture iodine and so on. Bromine is not used because of its high cost and toxicity. Iodophores are to be diluted in 50% alcohol while using for handwashing. To disinfect swimming pools, high concentration of chlorine is used. Contamination of spillage of all the infectious material can be treated with 0.5% sodium hypochlorite in virology and serology laboratories. Mercuric chloride is also used as a disinfectant in laboratories. It has staining and corrosive properties [17].

– Heavy metals

Heavy metals include mercury, silver, copper and so on. These have been used as disinfectants for long time, but at the same time, they are also toxic to environment. Heavy metals act on microbial proteins and results in the precipitation of proteins and hampers the activity of enzymes. The soluble salts of mercury, mercuric chlorides and so on are efficiently known for their bactericidal effect. Mercuric chloride and silver nitrate are widely used as 1:1000 aqueous solutions because mercuric chloride is highly toxic. In burn patients, silver sulfadiazine is applied topically to prevent bacterial colonization. Copper salts are heavy metals which are widely used as a fungicide [18].

– Surface active agents

Compounds that are fat soluble possess long-chain hydrocarbons, have charged ions which are water soluble that are popularly called as surface-active agents. These surface-active agents are effective against vegetative cells, few viruses and Mycobacteria. In hospitals, research centers and household, it is widely used as a disinfectant at a dilution of 1–2% [19].

– **Dyes**

There are two groups of dyes known as aniline dyes and acridine dyes which are used extensively as skin and wound antiseptics. Both of these dyes are bacteriostatic in high dilution but these dyes are of low bactericidal activity. Acridine dyes has the ability to act against nucleic acid of bacteria which makes it bactericidal. Dyes like malachite green, crystal violet and brilliant green are aniline dyes. Acriflavin and aminacrine are acridine dyes. A mixture of proflavine and euflavine is called as acriflavinea. Euflavine is the only dye which has chiefly effective antimicrobial properties. A related dye ethidium bromide (EtBr) that chiefly possess germicidal activity, intercalates between base pairs of DNA. They are more effective against gram-positive bacteria in comparison with gram-negative bacteria and are more bacteriostatic in action. To treat mild burn, these dyes may be used as antiseptics which may apply topically. In certain selective media, few of these dyes are used as selective agents [20].

– **Hydrogen peroxide ($H_2 O_2$)**

Since 1920s, hydrogen peroxide has been used as antiseptics. It has the property to kill microbes by acting on their cell wall. Hydrogen peroxide has the ability to produce hydroxyl-free radical that damages microbial proteins and DNA by the process of oxidation. It is widely used to disinfect the instruments. Approximately 6% concentration of hydrogen peroxide is used to decontaminate hospital equipment like ventilators. For skin disinfection, this solution at 3% concentration is used. Hydrogen peroxide solutions are also used against bacterial spores [21].

– **Ethylene oxide**

Ethylene oxide is a common gas used for the purpose of sterilization by attacking microbial proteins and nucleic acids. It is a compound which mainly acts as an alkylating agent. At room temperature, it becomes a colorless liquid. It has the strong ability to kill bacterial spores. It is mainly combined with CO_2 (10% CO_2 + 90% EO) or dichlorodifluoromethane because of its flammable property. It gets easily penetrated and absorbed by all the porous materials. Ethylene oxide is highly toxic, carcinogenic, mutagenic, explosive and irritating to eyes. Inanimate objects like petri dishes, syringes, bedding, rubber, plastics, dental equipment can be sterilized with the help of ethylene oxide. It has the ability to penetrate plastic tubes so they can be properly sterilized [22].

– **beta-Propioalactone**

At room temperature, most substances used for the purpose of sterilization do not have the ability to kill spores. beta-Propioalactone (BPL) is a colorless liquid which is formed by condensation of ketone and formaldehyde. It has broad-spectrum activity against many pathogens and spores that is why it is an effective microbicidal. To sterilize biological products like vaccines, enzymes, tissue grafts, BPL is widely used. In fumigation, BPL is more effective as compared to formaldehyde. BPL has properties like low penetrating power, irritant to eyes, carcinogenic and so on [23].

– **Acid and alkalies**

They are repressive to the expansion of the many microbes. *Mycobacterium* is more resistant against acid than alkalies. Boric acid is a weak antiseptic.

Table 9.2: Characteristics of main disinfectants.

Disinfectants	Bactericidal activity	Fungicidal activity	Virucidal activity	Sporicidal activity	Human toxicity	Applications
Formaldehyde	Very active	Very active	Very active	Less active	High	Surface and abiotic object disinfection
Glutaraldehyde	Very active	Very active	Very active	Very active	High	Abiotic object disinfection
Alcohols	Very active	Very active	Very active	Not active	Moderate	Skin disinfection, disinfecting small surfaces
Phenols	Very active	Very active	Less active	Not active	High	Inanimate object disinfection
Chlorhexidine	Less active against gram negative bacilli	Less active	Not active	Not active	Low	Skin and wound disinfection
Iodophores	Active	Less active	Active	Not active	Moderate	Wound and skin antisepsis
Chlorine compounds (hypochlorite, chloramine)	Very active	Active	Very active	Less active	Moderate	Water disinfection, skin, wound and surface disinfection
Hydrogen peroxide	Less active against cocci	Active	Active	Less active	Low	Wound antisepsis

9.8 Brief account of biomedical waste management

The importance of biomedical waste management and handling rules has been increasing for the past few years. Any kind of solid or liquid waste generated in hospitals or laboratories is called biomedical waste. This waste may be generated during research on an individual suspected to or suffering from any disease. Studies include short and long term research on observational, diagnostic, therapeutic and rehabilitative services [24]. Categories of biomedical waste are described in Table 9.3.

Table 9.3: Categories of biomedical waste.

Category	Type of waste	Type of bag	Treatment and disposal
Yellow	**Human anatomical waste**	Nonchlorinated yellow colored plastic bags	Incineration/deep burial
	Animal waste (It includes fluid, blood, tissues and organs of animals which are widely used in experimental and research studies. This kind of waste generated by animal houses of hospitals etc.)		Incineration/deep burial
	Soiled wastes (It includes infected body fluids including blood, cotton dressing, plaster casting etc.)		
	Discarded medicines and cytotoxic drugs (It includes waste and discarded medicines)	Nonchlorinated yellow colored plastic bags	
	Chemical waste (It includes chemical waste.)	Nonchlorinated yellow colored plastic bags	
	Liquid waste (It includes waste produced from disinfection activities from laboratories.)	Separate collection mechanism leading to effluent treatment system	
	Contaminated beddings and mattresses with blood or body fluids	Nonchlorinated yellow colored plastic bags	
	Microbiology and biotechnology waste (It includes wastes from laboratory cultures of microorganisms, human and animal cell culture, live or attenuated vaccines.)	Autoclave safe plastic bags.	Autoclaving/ microwaving/ incineration

Table 9.3 (continued)

Category	Type of waste	Type of bag	Treatment and disposal
Red	**Contaminated waste (recyclable)** (It includes waste generated from disposable items like catheter, gloves etc.)	Nonchlorinated red colored plastic bags	Autoclaving or microwaving followed by shredding.
White (Translucent)	**Waste sharps** (It includes waste from sharp objects like blades, scalpel, needles, syringes etc.)	Puncture, leak, tamper proof containers	Dry heat sterilization or autoclaving followed by shredding
Blue	a. **Glassware-** discarded and broken contaminated glass b. **Metallic body implants**	Cardboard boxes with blue colored marking	Disinfection

9.8.1 Waste treatment management

Choosing proper disposal methods is very important in waste treatment management [25–27].

The waste treatment and disposal techniques mentioned below may be applied for the treatment of hazardous medical waste. It totally depends upon the situation and type of biomedical waste generated for which we use different methods of disinfection and sterilization:

9.8.1.1 Thermal sterilization

– Incineration (200 °C to over 1,000 °C)

In this technique, waste is treated at a temperature more than 1,000 °C, which helps in its volume and weight reduction. However, such large-scale processing plants are designed for centralized networks, not for individual hospitals. Hospitals can use household refuse incineration plants that operate at around 850 °C. Medical waste is fed directly to kiln hopper. Blast furnaces or cement work incinerators can also be used temporarily but they are not usually recommended because the loading system of waste is not secured and also the emissions are not treated in such incinerators. Some simple incinerator models are available in the market which consists of single or combination of combustion chambers with a discharge chimney that securely removes the fumes generated from combustion. These incinerators are loaded with small sets of wastes that need to be incinerated at the intervals of 5–10 min after it has been lit for first half an hour. Generally, paper, wood or fuel is used to start combustion in incinerator. The incinerator is usually run at once for 2 h or more. While removing the ash, the personnel should wear heavy-duty gloves, body protection, goggles and respirator [28].

During this process of incineration, things to be kept in mind are:

1. Waste containing chlorine or PVC plastic should not be incinerated.
2. Maintenance of incinerator should be regular and faulty parts should be replaced.
3. Personnel operating incinerators should be well trained and possess operating manuals.
4. Emission control: Emissions must comply with BEP17 recommendations set forth in Stockholm convention.
5. Burning hazardous medical waste at dumps or in barrels must be avoided, due to the emission of toxic gases and also the leftover infectious materials, which can prove fatal to the staff.

– **Autoclaving (100 °C to 180 °C)**

It is a thermal process in which waste is subjected to a particular temperature (under pressurized steam) for a sufficient length of time (121 °C for 60 min) to be disinfected. Autoclaving is safe for the environment but it requires electricity, so it is not suitable for waste treatment in some regions. For sterilization of medical equipment, small autoclaves are used whereas for treating waste material relatively complicated and costly plants with internal mixing, shredding and drying systems are used. Massive autoclaves also need a boiler that generates various emissions that must be monitored. The effluents generated must be carefully disposed and monitored properly. After autoclaving, the waste can be land filled. Autoclaving is necessary for the infectious waste before it is disposed [30].

– **Microwaving (100 °C to 180 °C)**

It utilizes radiations produced by microwave to interrupt molecular chemical bonds and as a result disinfects waste. This method may need preshedding the waste first and then injecting it with steam treatment chamber. Thus, they heat it equally for 25 min at 97 °C to 100 °C. This disinfects waste without any harmful emission [31].

9.8.1.2 Surface sterilization and fumigation

Chemicals are generally used to wash off pathogenic microorganisms present in the medical equipment. These disinfectants generally have biocidal activities. This step generally removes contaminants from the surface hence, controlling their population to a level that they are not harmful. Surface sterilization and disinfection generally uses a bleaching agent which are sodium hypochlorite solution (1%), mercuric chloride (2–5%) or 0.5% active chlorine solution that releases chlorine which binds to ions in microbial cell and causes oxidative stress. These bleaching agents are generally used to disinfect patient wastes like urine, feces, blood and hospital sewage. Blood samples are generally mixed with nondiluted bleach for more than 12 h as they are rich in protein hence major source of contamination. Mixing of bleach with urine releases ammonia gas, which is toxic in nature so the waste has to be discharged into separate tanks with special provisions. These bleaching agents generally irritate eyes and respiratory system so proper precautions should be used before handling. Other disinfectants used are detergents like tween 20, sodium dodecyl sulfate that removes dirt and dust from instruments. From plethora of disinfectants, ammonium salts and peracetic acid are used to treat waste water. Ozone (0.5 mg/L) is used to wash of microbes from fruits and vegetables. Ozonated water generally reduces concentration of harmful disinfectants and hence limits resistance of pathogens to these chemicals. Solid wastes that need to be discarded are generally disinfected initially and then shredded or incinerated. As these chemicals are harmful and less effective, thermal sterilization is preferred over chemical disinfection [29].

9.8.1.3 Plasma technology

It allows an absolute and satisfactory destruction of waste. It reverses all hydrocarbonated goods into combustible gases without leaving behind any solid residues, without the need of waste segregation [32].

9.8.1.4 Needle extraction or destruction

Needles are removed from used syringes and destroyed, so that they cannot be reused and also to reduce the volume of sharps. These appliances run on electricity and destroy the needles by melting. These appliances need regular maintenance and should be handled with care. Plastic syringes also should be disinfected before they are disposed in plastic recycling [33].

9.8.1.5 Shredders

They cut the waste in smaller pieces. Shredders are often built into closed thermal or chemical disinfection systems. It requires well-trained staff for operating and maintaining this device. Only disinfected waste should be treated in shredders, so that they do not cause any harm to the staff. It is mainly used as a means of recycling plastics whenever syringes are available in large quantities [34].

Advantages/drawbacks
1. This leaves the waste unrecognizable.
2. This discourages the reuse of all the syringes and needles.
3. This diminishes the volume.
4. This encourages plastic recycling.
5. It improves the efficacy of chemical or thermal treatment in closed and integrated systems.
6. It requires electricity.
7. Some facilities in the shredder are very costly.
8. Large pieces of metal can cause damage to shredder.
9. If untreated waste is being shredded, the workers are exposed to air-borne pathogens.
10. Professional workers and continuous monitoring are required.

9.8.1.6 Encapsulation

Hazardous materials packed in containers made from nonreactive and impervious material is called as encapsulation. The coating materials are chemically stable, attached to the waste and resist biodegradation. The purpose of the treatment is to prevent humans and the environment from any kind of risk of contact. Biomedical waste are filled in the containers also requires addition of an immobilizing substance and properly seal the containers.

This process requires the use of metallic drums or the cubic boxes which are to be three-quarters filled with sharps, chemical or pharmaceutical residues. The containers or boxes are then filled up with high-density polyethylene and polybutadiene. When the used medium gets dried, then the containers are sealed and disposed in a waste burial pit and sanitary landfill.

The proportions which are recommended as per the guidelines of WHO is like 65% of pharmaceutical waste, 15% lime, 15% cement and 5% water. Encapsulation of sharps is generally not needed to be considered as a long-term solution. It is an inexpensive process [35].

Advantages/drawbacks
Biomedical waste can be disposed in a sanitary landfill. It is very essential that biomedical waste may be covered rapidly. The most popular method is to dig a pit down to the below level and dump the discarded healthcare waste immediately.

In the design and use of sanitary landfill, following are the important factors that must be taken into account:
1. Only controlled and restricted access should be provided;
2. All staff should be competent;

3. The discarding areas must be planned;
4. There should be waterproofed bottom of the landfill;
5. The water level must be 2 m or more, below the bottom of the landfill;
6. There must be no sources of drinking water or wells in the surrounding area of the site;
7. The waste must be covered properly to avoid vectors like insects, rodents and so on.

Before discarding hazardous medical wastes on the planned site, it must be inspected by water and habitat engineers to avoid any mishap [36, 37].

9.9 Education and training

For proper waste management in hospitals and research centers, education and awareness plays an important role. Appropriate education is required for all those who may come into contact with waste. Proper training and workshops are organized for all those who are directly or indirectly responsible for handling waste. All medical and related staff of hospital should be well aware of the protocols like segregation, storage and transportation. Within all medical facilities, sign and color coding should be extensively used. Overall responsibility for the management of medical waste should be given to senior member of medical or sanitation staff. Appropriate training in waste management should be given to the hospital staff which includes Doctors, nurses, laboratory staff, cleaners, hygiene promoters and medical support staff. For proper waste management, we should adopt following steps:
1. Appropriate waste management plan should be developed.
2. Use reusable products in place of single-use products.
3. In the patient's rooms, use small medical waste containers.
4. Always separate red bag waste containers from solid waste collection containers.
5. To ensure appropriate separation of wastes, color code containers must be used.
6. Make sure that wastes like chemotherapy and pharmaceuticals are being disposed properly.
7. Do regular checks to see if medical waste is being disposed correctly or not.

References

[1]   Culikova H, Polansky J, Bencko V, Hospital waste- the current and future treatment and disposal trends, Cent Eur J Public Health, 1995, 3, 199–201.
[2]   Appleton J, Ali M, Health Care or Health Risks. Risk from Health Care Wastes to the Poor, Leicestershire. UK, Water Engineering and Development Centre Lough borough University, 2000.

[3] Alzahrani MA, Alshanshouri MA, Fakhri ZI, Guide of Healthcare Waste Management, Riyadh (Saudi Arabia), Ministry of Health, 1998.

[4] Baldwin CL, Runkle RS, Biohazards symbol: Development of a biological hazards warning signal, Science, 1967, 158, 264–65.

[5] Kozlovac JP, Hawley RJ, Biological Toxins: Safety and Science, Biological Safety Principles and Practices, 2017, 15, 247–68.

[6] WHO. Laboratory Biosafetymanual. Geneva. World Health Organization 1993, 1–133.

[7] Basu RN, Issues involved in hospital waste management- an experience from a large teaching institution, J Acad Hosp Adm, 1995, 7, 79–83.

[8] Richmond JY, McKinney RW, Biosafety in Microbiological and Biomedical Laboratories, Washington, DC: U.S. Government Printing Office, 1999.

[9] Collins CH, Kennedy DA, Equipment-and Technique-related Hazards in Laboratory-acquired Infections: History, Incidence, Causes and Preventions, Oxford, U.K, Butterworth-Heinemann, 1999, 65–109.

[10] Draft guidelines on Hospital Waste Management issued by office of DGAFMS/DG-3A., New Delhi, 1999.

[11] WHO. Laboratory bio risk management strategic framework for action, 2012.

[12] Block SS, Chemical Disinfection in Hospitals, London, Public Health Laboratory Service, 1991.

[13] Gardner JF, Peel MM, Introduction to sterilization and disinfection, Churchill Livingstone, 1986.

[14] Rutala WA, Barbee SL, Aguiar NC, Sobsey MD, Weber DJ, Antimicrobial Activity of Home Disinfectants and Natural Products Against Potential Human Pathogens, Infection Control and Hospital Epidemiology, 2000, 21, 33–38.

[15] Moorer WR, Antiviral activity of alcohol for surface disinfection, Int J Dental Hygiene, 2003, 1, 138–42.

[16] Weber DJ, Barbee SL, Sobsey MD, Rutala WA, The effect of blood on the antiviral activity of sodium hypochlorite, a phenolic, and a quaternary ammonium compound, Infection Control and Hospital Epidemiology, 1999, 20, 821–27.

[17] Russell AD, Plasmids and Bacterial Resistance. Principles and Practice of Disinfection, Preservation and Sterilization, 3rd Ed., In Press, Oxford, England, Blackwell Science, 1992.

[18] Turnberg WL, Biohazardous Waste- Risk Assessment, Policy and Management, New York, John Wiley and Sons, 1996.

[19] Acharya DB, Singh M, The Book of Biomedical Waste Management, New Delhi, Minerva Press, 2000.

[20] Guidelines for Disinfection. Center for Disease Control, 2008.

[21] Sattar SA, Springthorpe VS, Rochon M, A product based on accelerated and stabilized hydrogen peroxide: Evidence for broad-spectrum germicidal activity, Can J Infect Control, 1998, 13, 123–30.

[22] Richard VS, Kenneth J, Ramaprabha P, Kirupakaran H, Chandy GM, Impact of introduction of sharps containers and of education programs on the pattern of needle stick injuries in a tertiary care centre in India, J Hosp Infect, 2001, 47, 163–65.

[23] Toplin I, Gaden EL, The chemical sterilization of liquid media with beta-propiolactone and ethylene oxide, J Biochem MicrobiolTechn, 1996, 311–23.

[24] Govt. of India. Ministry of Environment and Forests Gazette notification No 460 dated July, New Delhi, 1998, 10–20.

[25] National AIDS Control Organization. Manual of Hospital infection control, New Delhi 1998, 50–66.

[26] Mathur P, Patan S, Shobhawat S, Need of Biomedical Waste Management System in Hospitals - An Emerging issue, Curr World Environ, 2012, 7, 117–24.

[27] Gupte S, The Short Textbook of Medical Microbiology for Dental Students. Jaypee Publishers, New Delhi, 2012.

[28] Trulli E, Torretta V, Raboni M, Masi S, Incineration of pre-treated municipal solid waste (MSW) for energy co-generation in a non-densely populated area, Sustainability, 2013, 5, 5333–46.

[29] Park K, Text Book of Preventive and Social Medicine, Jabalpur, BanarsidasBhanot Publishers, 1997.

[30] Fleming D, Hunt D, Biological Safety Principles and Practices, Washington, ASM Press, 2006.

[31] World Health Organization, International Pharmaceutical Association, International Solid Waste Association. Guidelines for safe disposal of unwanted pharmaceuticals in and after emergencies. World Health Organization 1999.

[32] Coad A Managing medical wastes in developing countries: Report on consultation on medical waste management in development countries. InInter-Regional Consultation on Hospital/ Infectious Wastes Management in Development Countries 1994. WHO.

[33] PrussUstun A, Townend WK, Safe Management of Wastes from Health-care Activities, Geneva, World Health Organization, 1999.

[34] Reed RA, Dean PT, Recommended methods for the disposals of sanitary wastes from temporary field medical facilities, Disasters, 1994, 18, 355–67.

[35] Dubois M, Hoogmartens R, Van Passel S, Van Acker K, Vanderreydt I, Innovative market-based policy instruments for waste management: A case study on shredder residues in Belgium, Waste Manag Res, 2015, 33, 886–93.

[36] Gravers PD, Management of Hospital Wastes- An overview, Proceedings of National Workshop on Management of Hospital Waste, 1998, 16–18.

[37] Biswal S, Liquid biomedical waste management: An emerging concern for physicians, Muller J Med SciRes, 2013, 4, 99–106.

Index

https://doi.org/10.1515/9783110517736-010